On the Nature and Origin of Life

The History of Science
PREPARED UNDER THE GENERAL EDITORSHIP OF
DANIEL A. GREENBERG

ON THE NATURE AND ORIGIN OF LIFE

❁

Hilde S. Hein

McGraw-Hill Book Company
NEW YORK   ST. LOUIS   SAN FRANCISCO
DÜSSELDORF   LONDON   MEXICO
SYDNEY   TORONTO

Contents

On the Nature and Origin of Life

Life: Introduction

THERE are some things that everyone knows. We all recognize familiar objects and well-known faces, and we can distinguish those things that we know from one another. But sometimes we make mistakes, and sometimes looking at a thing from a novel point of view makes it appear very different. A seemingly solid and simple object such as a salt crystal or a strand of hair surprises us by its complexity when seen under a microscope. Things seen from unusual vantage points, either near or far, are often unrecognizably distorted. And those things which appear to us to be "as different as night from day" may turn out to be not so clearly differentiable after all.

Indeed, it is not altogether clear just how we do differentiate night from day. Is there a moment when the one leaves off and the other begins? Perhaps a meteorologist could tell us the precise time of sunrise, but does the dawn have an exact duration? Is it the fading out of night or the beginning of day? Or is there some kind of object or barrier which separates the two? If we cannot tell how they differ from or merge into one another, we cannot have as clear a notion of night and day or the difference between them as we thought we did. But what could be more familiar than night and day, and what could be more obviously different?

Similarly, we can recognize mountains and valleys and identify them on properly marked maps. But when you are out hiking or even flying over a landscape, can you really tell exactly where the mountain leaves off and the valley begins? To say we cannot differentiate between them is not to say there is no difference between them, but if we cannot state precisely what that difference is, then we must reconsider how well we really know what they are.

The same observation may be made not only of our knowledge or comprehension of things, but also of ideas. Some ideas are extremely familiar. We talk about love all the time; and we all know more or less what it is for people to love one another and even to love such a variety of things as one's dog, one's home, one's old hat, ice cream, the ballet, and swimming in the summertime. But do we have a very clear idea of what love is? Anyone who tries to explain it soon runs into trouble.

Similarly, we all have an understanding of what it means for a thing to be valuable. We compare objects to one another in terms of their value. We make value judgments of things and people, praising this one and denouncing that. But what are we identifying when we say that this is a valuable diamond and that a worthless piece of glass? Is value something to be touched or seen or measured, even indirectly as with a Geiger counter? What is the difference between the value of an object and all its other features—roundness or redness or juiciness—by means of which we identify it? Here, too, a very familiar notion turns out to be a mystery.

Among those ideas which we think of as commonplace is the concept of life. We know ourselves to be alive. Every infant rapidly learns to pick out the living things from the nonliving ones in his environment, but it would be difficult to say just how he does this. If asked how we knew that a thing was alive, we might answer that we saw it move, or that it acted on purpose. But other things besides living ones can move, and we can only guess that anything moves on purpose. In fact, we cannot clearly designate living things at all, and we cannot always distinguish them from nonliving ones. But this does not mean that we cannot learn or know anything about living things. Just as we need not know what love is in order to love somebody, so we are alive without knowing what life is, and we can know a good deal about living things without knowing what it is to be alive. Knowledge of living things —the kind of knowledge possessed by everyone to some extent and by the biologist professionally—is not the same as knowledge of life. To know the one is not necessarily to know the other. Some people are interested in acquiring knowledge and understanding of living things, while others are more interested in understanding the concept of life—what it is to be alive. Both are respectable problems: the first, a matter of observation and empirical research for the biologist; the second, a more theoretical inquiry and the concern of this book.

The concept of life may be studied from many points of view. It is not uncommon for teams of scientists, pursuing a problem from a number of vantage points and with a number of objectives, to pool their resources and information. They may be interested in different aspects of the same thing so that the same data serves several ends. Alternatively, inquiries may be carried on at different levels. The biologist takes it for granted that the creatures he studies are alive, and he asks by what mechanisms they maintain themselves, or perpetuate or reproduce themselves, or how they came to develop this or that evolutionary characteristic. He does not ask what it means for them to be alive in the first place. Some questions of this nature might be better left to the poet or artist or historian. They do not intrude upon one another, but it is difficult to pursue each of them simultaneously. The economist is interested in the value of a loaf of bread. He is not primarily interested in its taste, as is the baker, who seeks the finest flour in order to produce tasty bread. Only the housewife, who seeks the tastiest bread for the least cost, is to some extent assuming the attitudes of both economist and baker. Normally, specialists ask one kind of question to the exclusion of others.

Thus, the biologist does not ask what it means to be alive, and the more philosophically inclined theoretician does not concern himself with biological detail. He does ask about the concept: "What is life?" What are we saying of a thing when we say that it is alive? It should be noted here that the biologist does not have in mind the "meaning of life" in the sense in which theologians might use the expression. There is no reference intended to goals or purposes. The issue is not "Why are we alive?" (to what end?), but rather, "What is it that we are when we are alive?" The theoretician does not evaluate life or carry out practical experiments. He does not even need the magnifying glass which is indispensable to the naturalist. He engages in critical analysis, and he may do so from the comfort of his armchair.

Because the question "What is life?" is, to some extent, presupposed by questions regarding the nature and description of living things, it is sometimes believed that the theoretical or philosophical question is somehow more fundamental, closer to reality than the biological concern. But this is a misleading claim, and reflects biases about what is ultimately real, rather than stating a truth of any significance. Philosophical inquiry and biology may be carried on independently of one another, but the course of biology may be influenced by a theoretical formulation, and, conversely, theory may be affected by advances in empirical biology, or, for that matter, by the development of other sciences.

While the theoretician engages in no practical experimentation, he is interested in our experience of phenomena. Like the experimental scientist, he begins with common sense and ordinary experience; but while the practical scientist seeks to multiply experiences in a controlled fashion, thereby displaying their relations among themselves, the theoretician thinks about himself thinking about the experiences, thereby displaying his relation to them. He asks what it is that we know of the world when we have experience of it. He is interested in scientific theories as much for what they tell us about ourselves as for their utility in telling us about the object of scientific inquiry.

The practical scientist derives theories from his observation and experiments, together with certain beliefs that he holds. The philosophical theoretician is interested in elaborating those beliefs and making clear their relation to the theories formulated. While Newton's laws of motion are based upon observation, they also form a part of a theory about the nature of matter, and it is from this point of view that they are of interest to the theoretician. The doctrine of evolution is based in part upon the observations which Darwin made during his travels, but it also reflects his convictions regarding historical processes. Theoreticians study such theories not only as a guide to experience, but in order to understand what they entail, the premises upon which they rest, and the conclusions which follow from them. While they do not ask directly, "Is the theory true?" they consider what conditions would suffice to prove or disprove it. They want to know what a theory is, what its function is, and what, if any, bearing it has on facts and observations and objects in the world.

All theories are theories of something. They are explanations —some good, some bad, some narrow in scope, others far-ranging. But all of them purport to account for something, and they do this by relating concepts.

Concepts, unlike theories, are neither true nor false, good nor bad. Yet it is possible to study them. Practical scientists use concepts; theoreticians analyze them. Some people seek to evaluate concepts; others only explain and clarify them. Since theories relate concepts, a theory which uses muddled concepts is unlikely to be as successful as one which rests upon clear concepts. As long as our concept of person is obscure, we are unlikely to devise an adequate theory of personality. To this extent the philosophical analysis of concepts is prior to the formulation of scientific theories, although in fact the two intellectual enterprises cross-fertilize and stimulate one another. Many theories are weak because the fundamental concepts on which they are based are confused.

A theory of life would explain the phenomena of life, what those phenomena are, and how they came to be. It is not simply a description of living things, but an account of the life that they exhibit. Tremendous progress has been made in formulating such a theory of life, and it is the fruit of the joint effort of biologists, biochemists, geologists, biophysicists, and other scientists, such as astronomers, who also have something pertinent to contribute. The theoretician, too, has a crucial part to play in the exploration of the concept of life.

In the following chapters, we will be concerned with the theoretical analysis of life. We will not do any fieldwork or laboratory research or uncover any new factual data, although we will have to understand some of the substantive work that has been done in these areas. Our objective is a careful study of the concept of life. Our aim will be to achieve clarity, logical consistency, and faithfulness to the actual experience of life. A concept analyzed abstractly always runs the risk of being far removed from occasions for actual and practical employment. We shall try not to be too abstract or remote, but to remain as close as possible to our own and others' actual observations of life.

CHAPTER ONE

How Life Has Been Analyzed in the Past

The Nature of Life and the Origin of Life

MANY people believe that they understand a thing when they know how it began. This belief is so common that it has been given a special name, "the genetic fallacy." In fact it is not a fallacy at all, for we can learn a great deal about something by exploring its origins. Social scientists believe that knowledge of a person's beginnings and early environment give us an important clue to his later development and character. If you know how a curve ball left the hand of the pitcher, you can form a rather good estimate of where it will travel. Good baseball players and fans get to be expert judges of just such situations. Nonetheless, there is a difference between the nature of a thing and the beginning of that thing, and to tell how it began is not to tell all that might be known about what it is. For the sake of clarity, the study of the nature of a thing should be separate from the study of its origin.

But while the studies are separate, the thing which is studied is one and the same. It is that which *has* (or is acquiring) a certain nature which comes into being, and an account of its history must emphasize the singleness and continuity of its subject. A theory of the origin of life must be coherent with the theory of the nature of life upon which it rests. Clearly, some concept of life, however vague, is presupposed by any doctrine of the origin of life. For, in order to describe how something came to be, we must have some idea of what it is that came to be. It is thus extremely important that the two studies be coordinated. It would not do if I were to tell you that a cake expands in the oven because of the activation of yeast by heat, when, in fact, there was no yeast in the cake. Similarly, whatever we say about the origin of life and its development must be consistent with what is said about the nature of life. The concept of life as treated from both points of view must remain the same.

In the past this uniformity of concept has often been neglected, for the two lines of inquiry have generally been carried out independently. Experimental scientists have studied the origin of life in test tubes and have tried to project their studies to the primordial swamps; but the problem of the nature of life has a philosophical appeal and has been investigated as an "armchair" sport. Consequently, there have been great differences in the very language in which the two studies have been undertaken and very little attempt to reconcile the two.

The marriage between speculation about the beginnings of life and experimental attempts to initiate it was achieved by the Russian scientist A. I. Oparin. While Darwin had studied the evolution of living species, Oparin carried the investigation back to a period a billion years earlier, when the chemical elements which compose living matter were themselves evolving. In 1923, Oparin published a series of articles in Russian on the evolution of primary organic substances. This book, along with additional experimental substantiation of Oparin's ideas, became available in English in 1936; and it is only since that time that scientists have seriously endeavored to understand the nature of life by way of explicating its origin.

Oparin distinguishes between the formal questions "What is life?" and "How did life originate?" but he sees them as so related that we cannot answer the first without knowing the answer to the second. In taking this position, he is careful not to commit the genetic fallacy, for he does not claim to know a thing in its entirety by knowing how it began. Rather, he is maintaining that *all* phases of development are equally essential to the nature of a thing, and knowledge of all of them is necessary to knowledge of it. There is no more reason to concentrate upon the beginning than there is to single out the midpoint or the end as primary. The selection of any particular stage as that of "maturity" or "optimal development" is necessarily arbitrary and reflects the values of the selector rather than a feature given in nature. We do not, or should not, believe that understanding childhood is any less important to the understanding of human nature than is the comprehension of the adult. Similarly, according to Oparin and his followers, if we are to know the nature of life, we must account for its first appearance on Earth as well as all of its subsequent stages.

There is a common tendency to believe that living organisms, individually and as a whole, develop in the direction of a particular end or goal; that the caterpillar, for example, exists for the sake of the butterfly which emerges from its cocoon. But the caterpillar *is* the animal, not just because it has the capacity or potentiality of becoming the butterfly, but because the organism is all of those stages which make up a single unfolding life history. If we were to focus upon any one stage as common and crucial to all organisms, we might select decay and death as at least as universal as birth and so-called maturity.

We must, then, regard the questions "What is life?" and "How did life originate?" as intimately related, while granting that they may be distinguished in principle.

Life in General and Life as an Abstraction

We must also keep apart the notions of "life in general" and life as a property of an individual organism. We sometimes speak of life as if it were separable from living things, as if whatever we said of it were true of *it* and not of the things which are alive.

Similarly, we may say that power corrupts, meaning that people who have power are often corrupt, or rather, become increasingly corrupt as their power grows. We do not mean to say there is an independently existing thing, power, which operates on people much as oxygen causes iron to rust. There are only powerful people, and such people are often also corrupt.

So we may say of life in general that it exhibits one or another property, meaning by this expression that living organisms have certain properties. When, for example, we refer to the presence or absence of life on Mars, we mean to say that there may or may not be organisms which are alive on Mars. Their collective existence would entitle us to affirm that there is life on Mars; but this life (or lives) would not be life in general. Each organism is alive in its own individual way, and the life of each began in its own way, different from the manner in which life in general may first have appeared either on Mars or on Earth.

Our experience of life is limited to its occurrence in individuals on Earth. As yet we are not acquainted with any forms of life on Mars. Consequently, we must think of earthbound organisms as the model of life. When we speak of life in general, we are referring collectively to the existence and character of living earthly creatures; and when we speak of *the* origin of life, we do not mean that there is some shadowy substance, life, which is realized in, but is independent of the living things. What we do mean is that at some time, long ago, living things appeared on the Earth where earlier there had been none. There were living organisms, and, once they existed, there was life. (This, of course, assumes that life had a beginning, an assumption that needs investigation. We shall take up this question later.) To ask about the origin of life then is not to ask an abstract question, but is simply another way of asking concretely, "How did the first living things appear?"

Life as an Individual Occurrence

We must draw still another distinction between life as a generic phenomenon, which arises and develops within a number of organisms, and the mode of reproduction and development of those particular organisms. For while there is no life apart from that of the individual organism, we can talk about life as abstracted from the behavior of any given organism. Thus, while any organism which we might identify must have originated through the normal process of birth, we can talk intelligibly about the first appearance of life as independent of such processes.

Our inquiry will begin with the question of the nature of life, bearing in mind that we are making an artificial abstraction. Just as a sneeze may be studied as a dry academic object and from a position of hygienic safety (even though sneezes are in fact wet and occur only where there are noses and respiratory tracts), so life may be approached as a static and general topic, even though living things are dynamic and individual. Let us consider, then: What is the nature of life?

DEFINITION OF LIFE

Traditionally, people have thought that to study the nature of something is to give a definition of it. Definitions are supposed to do two things: they assign a meaning to a term, and they say something about that entity to which the term refers. Thus, if I say that a zebra is a black-and-white striped, four-footed mammal that lives in Africa, you can use the word "zebra" correctly, and you can also recognize or pick out that object to which the term is applied.

Because we come to know the nature of things by examining and studying them, many people have believed that definitions are discovered as the result of a successful search. The botanist, for example, seeks to understand what plants are by dissecting and comparing them. The task of science, then, is the revealing of definitions. But we are also taught that prior to any investigation there must be a definition of its subject matter, for one cannot inquire about what one does not know. How should we identify plants as such to begin with? We are faced with the dilemma that we need a definition in order to begin an inquiry, and yet the definition can only be the outcome of that inquiry. Clearly, this sense of definition is too rigid to be usefully employed.

In a sense, both horns of this dilemma state a truth, for we must have some idea of the subject matter we are studying in order to make further inquiry meaningful or even possible; and, on the other hand, we clarify that idea in the course of our investigation. As we find out more about things, we can speak of them with greater accuracy, and the resultant definitions will enable us to use the terms with greater success. This suggests that the meaning we assign to a term is closely related to what we think is true or important or interesting about the thing to which the term is applied, and this is, of course, subject to change as our information increases and our interests vary.

Since information and interests vary widely under different circumstances, so do definitions, and there is no basis, apart from the context of their use, for preferring one definition of a term to another. For this reason, we ought not to say that a definition is true or false, but rather that it is appropriate or inappropriate, useful or useless, adequate or inadequate, depending upon the circumstances.

If we consider the definition of the word life, it is evident that a variety of definitions might be appropriate, since, as we have seen, the topic may be approached from a number of points of view. From the vantage point of a child who needs to be able to differentiate living things from nonliving ones in his everyday experience, one definition will do, while a vastly different one would be proper for a biochemist who makes fine distinctions at the level of viruses or of complex protein compounds. Yet, since both definitions may well be adequate for their specific purpose, there is no reason to characterize one as more true than the other.

We may think of definition as a linguistic enterprise, having to do with the use of language for communication. As such, it is governed by linguistic conventions which are logically independent of the things to which the words in question refer. It is then possible to study those things and their behavior without first fixing upon a definition of the terms, and many scientists now prefer not to linger over definitions, but to deal directly with their subject matter. Provisional definitions may be used as a convenience, and some artificial concepts may be introduced and defined to serve as instruments in the investigations, but the formulation of definitions is no longer thought to be the primary task of science.

While the practicing biologist may successfully pursue his inquiry without defining life, one may well feel that the lack of an adequate definition is an intellectual gap. Philosophically or theoretically inclined biologists may still seek greater clarity in their analysis of the concept of life. They may extend our understanding of life by exploring our thoughts about it.

LIFE AND DEATH

A satisfactory analysis of the concept of life should enable us to distinguish between the living and the nonliving; and, further, within the category of the nonliving, to discriminate between the dead and the inanimate. Inanimate things are those of which it would not make sense to say that they are alive, while dead things are entities which once were, but are no longer alive. Only something which *could be* alive can be dead. The terms "life" and "death" (or "alive" and "dead") are so related that either both or neither are (successively) applicable to the same object.

This relationship has misled some people to believe that life may be defined in terms of death, *i.e.,* that living things are those which have a tendency toward death. But such a definition is fruitless; it does not permit one to go beyond what one already knows, for to know that a thing has a tendency toward death *is* to know that it is alive. It is as if I were to explain to you which side is my left by telling you that it is the one opposite my right. Obviously, if you were in a position to know the latter, you would also know the former.

It may appear as if one is learning a fact about the world when one is told that only those things which are capable of life are susceptible to death (or vice versa). But the connection is not made by studying dead or living things—and that is how we normally come to learn facts about them. Instead, one takes for granted the eventual death of a thing, or at least its loss of unique identity, when one identifies it as alive. More conclusively, one does not identify a thing as dead, as opposed to inanimate, unless one is convinced that it *has been* alive. There are circumstances under which one may be in doubt as to how to characterize a thing which is clearly lifeless; but that doubt would arise precisely out of one's inability to establish with certainty whether or not it underwent a prior living state. Once such evidence was found, we would have no doubt that the thing was dead.

This elaboration of the correlation between life and death is an instance of conceptual analysis, or intellectual clarification. It is not dependent upon the observation of living things, but rather requires that we think about what we mean when we call things living or dead. Such focusing of attention upon our own thinking often reveals relationships, as well as errors, which when confronted appear trivially obvious. But obviousness is not a mark of insignificance or lack of importance. Frequently in our search for the obscure we overlook those obvious and trivial things which turn out to be most important.

WHY ELABORATE THE CONCEPT OF LIFE?

One of the seemingly obvious, but in no way trivial demands that must be made of an adequate definition or analysis of life is that it be compatible with those features which we have observed to be characteristic of living things. It cannot contradict anything that we know to be true about living things. It is in this respect limited by what we do know, and it may serve as a restraint on the acquisition of further knowledge; but it may also stimulate the expansion of scientific knowledge. Sometimes the mere restatement of ideas clearly and in logical order is sufficient to bring to the surface implications which had not been previously evident.

But while a concept of life must be consistent with all that we know to be true of living things, it should not be merely a summary statement of these known truths. What we expect of a concept of life is that it provide us with a *criterion or set of criteria* for the identification of life. This is to say that we know that a thing is alive when we know that certain conditions have been fulfilled. We do not have to know everything that is true about living things in order to recognize them as living. The question is, "What do we have to know?" And that is a difficult one to answer.

Sometimes our information about what makes a thing what it is is absolutely conclusive. So, for example, if *x* is known to be a three-sided enclosed plane figure, then it follows necessarily that it is a triangle; or, if John is known to be my father's brother, then it follows necessarily that he is my uncle. But sometimes our reasons for saying that *x* is a thing of a certain kind are less positive. Suppose, for example that I know only that my grandfather's will provided for his money to be equally divided between John and my father. It certainly does not follow with any necessity that John is my uncle, but this might be a pretty good reason for thinking he is. We decide upon good and bad reasons of this type on the basis of experience; and, in the context within which they are reasons, we regard them as evidence for or against certain conclusions.

In the case of the triangle, we have an airtight case. We know that anything which is a three-sided enclosed planc figure is a triangle, just as anyone who is my father's brother must be my uncle. That is, he fits the qualifications or fills the criteria of being my uncle. Through study and observation we can come to know which of the things that we experience possess the qualifications to be things of a certain kind; but to decide what these qualifications are is a matter of reasoning and understanding. Our problem is to establish what qualifications a thing must possess in order to be alive. When do we say that the criteria of life are satisfied?

We might say "I'm sure that x is alive, because it just bit me," and in most circumstances that would be convincing. Most things that bite are alive; but we can think of things, such as bear traps and false teeth, which could bite and yet are not alive. Furthermore, the test would not always work, for some things which we would characterize as alive cannot bite. So while the ability to bite might constitute reasonably good evidence that a thing is alive, it does not serve as an adequate criterion of life. An adequate criterion would permit us to make the immediate step from the recognition of its satisfaction to the awareness that x is alive. To recognize the one is to know the other.

Once armed with a criterion of life, we can go on to investigate the conditions under which life is possible, the causes which bring it about, and the properties which are characteristic of it. But it should now be clear that these conditions, causes, and properties are not what we mean by life. They are not criteria of life, although they may be used as evidence that a thing is alive. A practicing biologist tends to concentrate on empirical features. His job is to elaborate the physical properties of living things, to show how they arise, and how they are related. Thus he studies such characteristic attributes of living things as reproduction, metabolism, sensitivity, respiration, and regeneration, and he may come to think of them as definitive of life. But, in fact, his selection of these as characteristic features of living things presupposes his knowledge of what it is to be a living thing. It is taken for granted that he can identify a living thing, and no doubt he can. But he does not spend his time articulating the method by which he makes that identification. That is a job for the theoretician. As we have seen, the two tasks are closely related and may limit one another. Historically they have often been confused. Biologists, mistaken in the belief that any practical investigation must be preceded by a definition of the object investigated, have fixed upon one or another empirical property of living things and identified it as the "essential" or defining property of life. But such designations have always been disappointing; for invariably a detractor unearths an example of a clearly living thing that lacks the "essential" property, or someone else contrives some obviously mechanical system that exhibits it. Fortunately, we may be consoled by the thought that the progress of practical science depends far less upon the amplification of theory than upon careful experimental work within the confines of what is known.

One of the curious by-products of the separation between the theoretical and the practical approaches to life and of the problem of defining life has been the hardening of ideological lines around two schools identified as *vitalism* and *mechanism*.

The dispute between these two positions cannot be expressed as a simple disagreement over biological fact or doctrine. It has come to represent opposing life styles and ways of looking at the world; but clearly one of the central issues which divide mechanists and vitalists is that of the proper analysis of life.

No single analysis of life is adhered to by all members of either school, and a variety of analyses is compatible with each. It is, however, characteristic of mechanists to regard life as a natural phenomenon, possibly more complex but essentially similar to and continuous with ordinary nonliving natural events. Vitalists tend to think of life as a unique event, possibly wholly inexplicable, and in any case requiring special principles of explanation to account for the difference between the living and the nonliving. Where a special explanatory principle is introduced to account for a set of phenomena which appears to be extraordinary in terms of accepted theory, we refer to that explanation as "discontinuous." Sometimes the discontinuity is a matter of scale. Thus a landslide, or avalanche, involves no physical actions which do not occur ordinarily as a part of normal erosion and sedimentation. But because of the concentration and intensity of these events under some circumstances, one tends to segregate them as a single, detached event. Vitalists tend to emphasize and dramatize such "detached" events of nature and the need for discontinuous explanations of them, while mechanists stress continuity and seek to establish continuities where they are not evident.

Let us turn now to some of the conceptual analyses of life which have been proposed in the past and try to understand what they entail. We shall begin with the most extreme vitalistic (that is, discontinuous) formulations, and move on to positions of a more mechanistic character.

CHAPTER TWO

Vitalistic Theories

Theism

A THEIST is a person who believes in the existence of God or gods. As such, he need not hold any particular views regarding the nature of life and its origin; nonetheless, it is commonly expected that theists are also vitalists. This is not a necessary expectation, but it is also not an unreasonable one. In line with our earlier remarks about continuous and discontinuous explanations, we may note that theists generally accept the discontinuity of the universe. Having admitted that God exists and is responsible for a number of otherwise inexplicable features of the universe, one has no grounds for limiting the extent of phenomena for which God might be responsible. God is easily invoked whenever our understanding is frustrated. Thus one may readily attribute to God the creation and maintenance of life, thereby designating it an event which transcends the boundaries of ordinary physical explanation.

The predisposition to accept discontinuity in the form of divine intervention at any level of experience may also predispose one to be a vitalist. Similarly, if one believes that disease and madness are the result of spiritual possession, one may be ready to attribute a similar cause to artistic creativity. If, on the other hand, one insists that disease has purely organic causes, one is less likely to

accept an explanation of artistic genius in terms of divine or de-
monic inspiration. The connection between one form of discontinuity and another comes from psychological associations,
and itself forms the basis of a new continuity of belief, namely, a
readiness to accept events as novel or inexplicable. Thus it frequently turns out that theists are also vitalists, but the reverse is
not necessarily true.

Theism is also compatible with mechanism. One of the greatest
European philosophers, René Descartes (1596–1650), who has
been regarded as the father of both modern science and modern
philosophy, was a mechanist and a theist. While God plays a fundamental role in the formulation of Descartes' philosophy, he explains life as a purely mechanical operation of a highly complex
physical system. Organisms are simply complicated automata.

It is also possible to be a vitalist and a nontheist, as well as a
mechanist and a nontheist. In principle, then, the issue of theism
and nontheism is wholly independent of the mechanism-vitalism
controversy, and the association is mentioned here only in order to
correct the popular misconception.

One form of vitalism, however, does make use of a particular
theistic principle. This is the doctrine of the "divine spark." It is
maintained that the breath of life or spark of divinity is implanted
in the organism by the Creator. Literally, this is inspiration, and
the evidence in support of the event is frequently drawn from the
Bible or other scriptural sources. Since the event is of a supernatural quality, there can be no direct empirical evidence for it. Furthermore, it is not clear whether all living creatures possess a divine spark or whether this is a privilege reserved for man. The
doctrine of Genesis is compatible with the belief that not all living
creatures were directly created by God, since some of the lower
organisms may have arisen spontaneously out of the dust and
slime of the earth. Something other than sheer life must be the
mark of divinity, since life is a property common to man and
other organisms, however they are characterized.

Mechanists generally reject the thesis that life is identical with a divine spark. They hold that the doctrine is obscure and misleading since there is no relevant evidence which could serve to deny or substantiate it. We do, of course, have evidence that a thing is alive. But nothing is gained by regarding this same evidence as testifying to something beyond those signs of life. We could say nothing more of the shadowy divine spark than that it *causes* those very features which we identify with life. It is as if I were to insist that the cold from which I suffer is something apart from all the coughing and sniffling and wheezing which you observe, these being merely effects of the cold. Normally we are satisfied that these events and characteristics *are* the cold and are not merely evidence of it. This is not to say, of course, that there is not something, a virus, which causes the cold. But the virus is not identical with the cold, which is its effect.

It is, furthermore, not the case that we are never justified in accepting one thing as evidence of another. A hole in the skin of my apple may lead me to suspect that a worm is, or was, at the core, and I can then cut open the apple and see if there is one. But suppose I said the hole in the apple skin was evidence of an apple-skin borer which could not be seen or touched and which could not be described at all except by the statement that it bored holes in the skins of apples. Would my telling you that the hole had been bored by such a creature tell you anything more than that there was a hole in the skin of the apple?

This is the predicament we are in with respect to the divine-spark explanation of life. For there is no way to characterize the divine spark except as that which produces life. Hence, having described the object as alive, I have said all there is to say about the spark. There is no independent way even of establishing its existence. The vitalist who attributes a divine spark to the living organism is not providing an explanation of life, but is in fact merely calling attention, in obscure language, to that remarkable phenomenon of life which calls for explanation.

Mechanists reject such an account as not merely false, but as meaningless. Unless some independent characterization of the divine spark can be delivered, there is no reason to speak of it rather than of the life features which constitute the evidence for it. Indeed, references to the divine spark have the effect of lulling people into the false satisfaction that they have made an important discovery, when in fact they have merely reasserted that there is something to be sought.

The vitalist might respond here that no burden of proof falls upon him. He is attributing a causal role to the divine spark, a supernatural agency whose consequences may be experienced but which is itself beyond empirical inquiry. We do have empirical evidence that living and nonliving things differ from each other, and so, he will say, we must look for a cause of this difference. And, failing to find such a cause in the empirical realm of natural phenomena, we must look for it in the nonempirical realm of supernatural phenomena. Notice that the vitalist is reasoning along wholly intelligible causal lines, exactly the same as the mechanist. For every event there must be a cause, and wherever there is a variation in events, there must be a variation in causes. The mechanist, however, restricts his search for causes to the natural domain. Where he fails to find causes, he rests content to describe the inexplicable phenomenon, hoping that eventually a causal explanation continuous with others will turn up. The vitalist, meeting the same barrier, leaps across it and finds a nonnatural or supernatural "cause" of the event. But this is an odd sort of conclusion to draw from the absence of causal evidence.

If we have no evidence of a cause of an event x (which we nonetheless know to have occurred), then we are not justified in claiming that nonetheless there is a unique sort of noncause (or supercause) which is responsible for x. We are entitled only to deny knowledge of the cause. Negative evidence is the *absence* of evidence; it is not a positive form of nonevidence or a form of positive support for something negative (for example, a noncause). Where we have negative evidence, we have no warrant for drawing any conclusions whatever.

The vitalist who uses negative evidence to draw positive conclusions is playing two games simultaneously. Agreeing with the mechanist that no natural explanation of life has thus far been found, he abandons the mechanist's search and treats that very absence of evidence as positive grounds for affirming some supernatural explanation of life. But no further evidence is relevant here, for the supernatural is by definition beyond the natural, outside empirical inquiry. Arguments for supernatural causes cannot be refuted by empirical evidence to the contrary; no more can they be established by empirical evidence. Thus the objection raised against the vitalist is not that he believes in supernatural causes (no denial is made of the supernatural). It is rather that he pretends to defend the supernatural with arguments appropriate only to the natural realm. The mechanist would be equally at fault if he were to support naturalistic conclusions by appeals to mystic revelation or to other nonnatural sources.

Theism as such may well be true, and, as we have seen, it is compatible with both mechanism (*e.g.,* Descartes) and vitalism. Our concern is neither to establish nor to deny theism, but only to point out that its truth or falsity cannot be determined on empirical grounds. Hence a vitalistic position which rests entirely on the premise of theism cannot have the same logical form as a mechanistic doctrine based upon the continuity of naturalistic explanation. Consequently, since the common ground of argument has been abandoned, there cannot be a reasonable dispute.

Our interest in mechanism and vitalism is focused upon their disagreement over the nature and origin of life. Hence, for the remainder of this book we shall set aside controversies which do not permit dispute. (If you like lemon in your tea, and I like milk, we may think one another a barbarian, but there is really no intelligent way in which we might persuade one another to think otherwise.) Let us turn, then, to another vitalistic analysis of life, one which can meet the charges of mechanism on its own grounds.

The Vital Principle

Of all the possible forms of vitalism, the one which is most popular and which is generally meant when anyone speaks of vitalism is the doctrine of the "vital force." According to this view, living things are differentiated from nonliving ones by the presence of a special force or energy. This unique "life quality" is passed on from parent to offspring, but is never found in nonliving entities. There are a number of variations upon this thesis.

From antiquity to the present, it has been believed that only living things have a soul or *anima,* sometimes identified with mind, which varies from organism to organism. A plant has a less-complex soul than an animal, for while it does take in food which it uses for growth and reproduction, it does not do anything so complex as to move about extensively or to interact *perceptually* with its environment. Soul may be represented as if it were arranged on a scale starting from the lowest animate creatures (perhaps some form of lichen or mold), which are limited by their material form, and ascending to more-complicated animals and men, and then ultimately to purely spiritual beings which exist without any dependence upon matter.

The body which possesses a soul can be understood as a material object which acts in accordance with the known laws of nature. But, if the soul or mind is immaterial, and if it can be understood by human reason at all, it must be investigated by methods of a totally different kind. For while the soul is a substance, it is a substance of a different and more elusive nature than the body. Sometimes the soul or vital substance was thought to be a very delicate stuff which would be easily destroyed if its body or material container were given rough treatment. But it was often thought to outlast one or several specific material containers.

One man who believed in a special, strong but sensitive, vital element was John Needham (1713–1781), an English naturalist and theologian. In 1748, he performed an experiment in which he boiled mutton broth in order to kill all the organisms contained therein, and sealed it in glass containers. A few days later he opened the containers and found that living organisms were present. Needham claimed that the sensitive vital element continued to reside in the boiled broth, although the living organisms had been killed, and that it had produced new organisms, such as maggots. Needham was a vitalist who believed that chemically organic, but nonliving matter could be organized into living organisms by an imperceptible vital element. His experiment was designed to prove that living organisms could be caused to arise from nonliving matter by the action of the vital element. However, twenty years later, Lazzaro Spallanzani (1729–1799), an Italian biologist, repeated the experiment. He boiled the solution for a much longer period of time with the result that no microorganisms appeared, even though the solution stood for a longer period of time. Spallanzani concluded that Needham had simply not boiled the mutton long enough to kill all the organisms. Needham rejected Spallanzani's conclusions, saying that Spallanzani had destroyed the vital element along with the individual organisms by boiling the solution too long.

While Needham and others believed the vital element to be a special substance, some vitalists concentrated instead on characteristic activities which they believed to be exhibited exclusively by living creatures. Albrecht von Haller, an eighteenth-century Swiss physiologist, believed that the organs of living beings exhibit certain vital activities, such as contraction, sensibility, and irritability. He observed that a slight stimulus could produce a movement which was altogether out of proportion to its mechanical disturbance, and that this tendency persisted only so long as the organism was alive. Once the organism was dead, it behaved mechanically (proportionately to the force applied to it), like any other inanimate object. Von Haller concluded that there was a unique vital property which disappeared when the organism died.

Further support for this view existed in the fact that there were indeed different chemical compounds existing in living and nonliving entities. The now classic division between organic and inorganic chemistry stems from the belief that there are some chemical compounds that are manufactured only within living organisms. The elements of which they are composed might be the same as those of inorganic compounds, but their particular organization was unique and had profound effects which could only be realized by the force of some vital principle.

In 1828, a major event took place. The substance *urea* was a well-known organic compound which is found in the blood and urine of mammals. It is regularly formed in animal bodies in the process of breaking down protein. Urea, a typical animal waste product, was generally believed to be obtainable only as manufactured within an organism by means of a vital force. But in 1828, a German scientist, Friedrich Wöhler (1800–1882), synthesized urea in a laboratory out of ammonium chloride and lead cyanate. Other organic syntheses soon followed.

The synthesizing of these compounds led to the conclusion that while differences might exist between organic and inorganic compounds, there was no significant difference in their method of production. Substances which scientists had formerly believed to be produced only in organisms under the influence of a vital force could now be artificially produced in a laboratory. This discovery did not completely put an end to vitalism, but it did undermine the doctrine that organisms are composed of a substance which is fundamentally different in nature from the matter of inorganic objects. Physiological activities were now more commonly explained in terms of ordinary physical-chemical processes. The vital force was no longer invoked to explain the contraction of muscle or the activity of nerves. Digestion and nutrition came to be regarded as the source of chemical energy which keeps an organism going, and this activity does not require the presence of a special vital principle.

Still, this revision of views is not incompatible with some faith in a vital principle. Many people believed that while there was no special vital substance, and while the vital force was not directly responsible for organic activities, it was still necessary as a guiding and sustaining principle. The complexities of organic behavior were so remarkable that they required something more than purely physical arrangements and chemical attractions.

Increasingly, in the course of the nineteenth century and the beginning of the twentieth, the significance of vital-force theories declined. As the experimental evidence mounted, it became evident that many of those phenomena previously thought to be vital and unique could be explained in nonvitalistic terms. Even the law of conservation of energy was discovered to apply to living as well as to nonliving processes. It was clear that wherever there was a production of power, there was a corresponding exhaustion of something to supply it. Organisms, just like the fuel engine of an automobile, are dependent upon an external source of energy. Ultimately this source is the sun.

It has been claimed that the doctrine of evolution had something to do with the demise of vital-principle theories, for it made possible the explanation of the coming into existence of new and increasingly complex kinds of living creatures. But, in fact, there is nothing in the doctrine of evolution which would preclude the existence of a vital force. Indeed, some people believed that evolution itself was propelled by a vital force.

The playwright George Bernard Shaw (1856–1950) held such a view of evolution. In a long introduction to his play *Back to Methuselah,* he criticized the Darwinian theory of natural selection as "a chapter of accidents." Such a doctrine was morally offensive to Shaw, and he preferred to believe that there was an initial creative impulse, or vital force which expressed itself through evolutionary change. Life advances to a stage of perfection by a trial-and-error process of self-correction. Man occupies a high place along this evolutionary scale, but he is not yet perfect; and, Shaw warned, if he does not pull himself together and make a success of life, man will be displaced by a higher species, one which will profit from man's experience.

The philosopher Henri Bergson (1859–1941) also held a view of evolution dictated by a vital force. In 1913, he wrote a book, *Creative Evolution,* which to many people has served as the model of a philosophic defense of vitalism. Bergson regards the life force, or *élan vital,* as a kind of current which passes through matter, organizing it into individuals of ever-increasing complexity. Chief among its characteristics are the creative capacity to generate novelty and, along with this, a basic indeterminacy or unpredictability. These are the features which distinguish living things from inanimate ones.

Matter may be organized in a variety of ways, and man has acquired the skill of building extremely complicated systems, many of which exhibit some of the patterns of interaction and self-regulation of living organisms. Think of robots and computers, of thermostatically controlled heating systems, of automatic transportation complexes, and of the information-processing devices of systems engineers. All these processes which have been traditionally associated with living creatures can in fact be synthetically duplicated in obviously nonliving structures. Clearly, then, it is not the processes in which organisms engage that determine whether or not they are alive. These processes are not only explainable in purely physical terms, but can be simulated in machines which no one would for a moment consider to be living. Does design make the difference? Does the difference lie in the fact that the machine is the result of a conscious human plan, while the organism is the product of natural reproduction? This account of their distinction is unsatisfactory for a number of reasons.

First of all, it presumes to explain the differences in the nature of things in terms of their origin. But that explanation commits the genetic fallacy, and surely it is not merely the conditions of their creation that we have in mind when we differentiate a machine from an organism. Secondly, it suggests that the mode of origin may be the *only* difference between the living and the nonliving, and thus that if an object resembling an organism in all other respects were in fact to be synthesized in a test tube, we should be forced to classify it as nonliving. Conversely, if techniques of automation were to become sufficiently perfected to permit machine production of machines without any intelligent human intervention, then we should have to acknowledge that such mechanical offspring were in fact alive.

But the difference between living and nonliving things is not in how they came into being, but rather in what they are. According to Bergson, this difference reveals itself in the impossibility of the intellect's truly understanding the nature of living things. The intellect is adapted to the static organization of matter. This is artificial and predictable. If you know the principles of physics well enough you can take a physical system and predict how it will behave under any given conditions. If somehow it worked out to surprise you, you would blame that on your ignorance. But in the case of the living organism, one is constantly surprised. New and unexpected things appear without warning, and they cannot be dismissed as reflecting our ignorance, for the unpredictability is built in. Life cannot be analyzed in rigid terms, for it is an ongoing process, flowing through time, ever changing, producing novelty, literally evolving creatively.

It is important to note that Bergson does not believe in a fixed final point toward which evolution progresses. This would make life a mere process which selects "the best" among alternate means to achieve a certain end. Life has greater freedom. Living beings are in fact independent centers of activity which organize matter about themselves in accordance with their own perceptions. Each organism shapes its environment according to its needs. No two snail shells are exactly identical. Nor is your home exactly like that of your neighbor. Each reflects the needs and efforts of its inhabitant.

Living beings are constantly creating new fields of activity about themselves, new bases for self-development. Thus, in a sense, even the material world is a product of living evolution, rather than the reverse. Life is not something super-added to matter, an afterthought. Life itself is the force which impels matter to organize itself into ever-expanding combinations. To describe these combinations from the outside is to fail to comprehend the "ongoing" character, or inner fluidity which is the real essence of life. Life cannot be intellectualized; it can only be intuited from the inside and lived through.

What reasons does Bergson give for characterizing life as he does? Is there any evidence in support of his view other than the failure of his opponents to adequately explain the difference between living and nonliving organisms?

Most of Bergson's arguments are negative. He opposes both mechanism and neovitalism, calling them finalism, because they are counterintuitive. Our inward experience of life and of duration, rather than segmented time, is that of a stream which is ever expanding, ever creative. We cannot, he says "sacrifice experience to the requirements of a system." Any system which entails a preordained plan toward which organisms move is necessarily false, and all systems, being works of the intellect, necessarily impose such a plan. In effect, then, Bergson is saying not merely that all existing explanations of life are contrary to experience, but that such explanations must be inadequate insofar as they are intellectualized. The experience of life is by definition nonintellectualizable. The problem is not that at this point in history scientists have failed to come up with a satisfactory account of life, but that no satisfactory account can be given. And how does Bergson know this? By experience.

This is a rather peculiar argument from experience. One may feel dissatisfied with a rational account of something and believe that since the account does not explain all features of experience, a better account may yet be discovered. But how can one have an experience of impossibility? How can one discover through experience that any rational account is impossible? Is this not very much like Alice's meeting Nobody along the road?

It is possible to provide stronger evidence based on experience in favor of the vital principle, and sometimes Bergson appears to offer that evidence. Some people have called attention to a kind of urge or tendency which appears to characterize living things. If you search within yourself, you will become conscious of a vague striving which is always present. We cannot, of course, be absolutely certain that the same "urge" is present in other creatures, or that it is absent in inanimate objects. But it seems reasonable to attribute the same kind of experience to at least the higher organisms and to deny it to objects whose motion is clearly a consequence of external forces. Bergson finds further evidence of this vague striving in the plurality of lines of development which evolution may take. The variability and adaptability of life forms are, first of all, taken as evidence of the absence of a determined plan, assuming that such a plan would provide for efficiency and economy. Secondly, plurality of life forms is taken to imply the presence of a positive multipotentiality inherent in the life force. Bergson is not satisfied to deny the existence of a program; he regards this very nonexistence as a sign of a positive (nonprogrammatic) feature of living things.

It is not altogether clear how the absence of definition of life can be interpreted positively as the presence of a plurality of potential definitions. One can, of course, point to the many lines of development which life has taken, and this is what Bergson does. Assuming that life had a common origin, it is remarkable how many distinct lines of development have taken place—some of them successful, and some not. The most noteworthy ones which Bergson discusses are the development of instinct in the insects, and of intelligence in man. Bergson entertains the possibility that one day there will be a creature which unites the advantages of both faculties, although now they are seen to be mutually incompatible. Such a creature would be supremely well endowed, but, according to Bergson, this creature may never come into being. Since life is unpredictable, its evolution cannot be taken as a certainty.

Yet hidden in Bergson's notion of evolution is the idea of progress. The various forms which life takes are not only diverse, they are increasingly complex. The development of the nervous system implies increased consciousness, which in turn provides increased opportunities for choice, and hence increased freedom. Insofar as freedom is a condition of inventiveness and of greater indeterminacy, the evolution of the vital principle is a self-generating, expanding stream. Bergson's vital principle is identical with the great "river of life" which flows through matter, diverting it to the production of more and more highly evolved life. While its final realization is yet to be achieved, its highest production so far is the development of mankind.

Bergson explicitly rejects finalism, or the contention that evolution is directed toward the achievement of a specific preordained goal. Finalism, like mechanism, he says, denies the role of vital freedom. Both reduce life to a mechanical coordination of means to ends, and thereby overlook the inventive creativity that is the true character of life. Nonetheless, he does regard life as evolving in a linear direction. Like Shaw, he believes in a form of progressivism, according to which life moves toward ever-greater perfection. While this progression is not ordained and inevitable, it is clearly represented as being "in the nature of things." It is perhaps this optimism as well as the striking imagery of Bergson's language which has given this particular form of vitalism such popular appeal.

Entelechies

Almost as popular as the vital-force theory has been another form of vitalism which also has an ancient tradition but which appeared in its modern form as an alternative to vital-force theories. This is the doctrine of *entelechies,* made famous in modern biological history by the German biologist, Hans Driesch (1867–1941).

Driesch's objections to the vital force were not directed against the beliefs of Bergson, but against those of another great contemporary scientist, Hans Ostwald. Ostwald believed that the vital force was a type of elemental energy, analogous to electrical energy and responsible for biological phenomena and activity. Driesch objected to this analogy on the ground that life is something qualitative, not quantitative, and that vital energy, if there were such a thing, would therefore be nonmeasurable and nonstorable, unlike ordinary forms of energy.

Nonetheless, Driesch agreed with the proponents of the vital force—that life is a unique event and that biology is therefore a science independent of physics and chemistry. Life, he says, does not just result from a group of complicated inorganic conditions. It is not a kind of stuff; nor is it an elemental form of energy, as Ostwald believed. Yet it must be "a true element of nature," ultimate and irreducible. What then is the vital factor?

Driesch borrows the term entelechy from Aristotle, whom he credits with being the "founder of theoretical biology" and the "first vitalist in history," but he gives the word a new significance. Aristotle identified the entelechy with the *soul* of a living organism. He regarded it as the essence or guiding formula of the entire living being, the principle which determines that a creature will develop to be an organism of a specific kind. Thus the factor which makes a puppy grow up to be a dog, and not a giraffe, is its entelechy. This factor must be actually "contained in" the puppy at the outset, as a goal toward which the organism strives. The entelechy guides its development. In effect, the entelechy is naturally present within the organism in the same way that a plan is present in the mind of a designer when he creates an object—whether it be a house, a dress, or a machine. The primary difference between living things and organized nonliving things like houses is the fact that in the case of a house, an *external* principle, the architect's drawing, determines its growth and development. In the case of organisms (living things), this principle is internalized as the soul, or entelechy. It is important to note that, for Aristotle, the entelechy is not a substance which may exist independently of the body to which it gives form. It is not a distinct, immortal soul. As the form *of* the body, it has no reality apart *from* the body. But we can talk about it separately, as we can speak of the form of a sonnet, or the correct form of a golf stroke. However, the entelechy is not simply an abstraction. It actually has an active role, causing the organism to reach the state of maturity.

Driesch assigns a more modest role to the entelechy. The determination of what sort of thing an organism will grow up to be is fixed by its physical nature. A puppy could not be anything other than a dog because of its material constitution. However, there might be injuries or early damage which interfere with normal development and which require adaptation so that a mature dog can be formed despite conflicting circumstances. These adjustments are controlled by the entelechy.

Driesch's views were based upon his own experimental research on the development of animal embryos. He was particularly interested in morphology, the study of the structural development of organisms. He was struck by the flexibility which organisms show in the earliest phases of their development. Normally a regular pattern of growth occurs, and each portion of the original cell follows a set procedure which makes a specific contribution to the development of the whole. But if this procedure is interrupted through injuries or other abnormalities, then, Driesch discovered, the normal pattern of development is altered in such a manner as to compensate, or to assure the development of the whole by way of another route. Driesch experimented with sea-urchin eggs, which he divided into two or more portions at a very early phase of their growth. To his surprise, he found that instead of forming half an urchin embryo, as he expected on the supposition that each portion of the original has one specific task to fulfill, each half formed one complete, but undersized sea-urchin embryo. This discovery and others like it led Driesch to question the variety of routes which an organism may take in order to preserve its integrity as an organism. He pursued similar investigations concerning the capacity of some organisms to regenerate lost parts. This ability also is an instance of an organism's overcoming adverse circumstances in order to maintain its identity as an integrated whole. Unlike an incoherent collection of parts, the parts of an organism appear to "belong" together. Driesch wondered if this phenomenon could be explained in purely mechanistic terms, and he concluded that it could not.

Driesch was struck not merely by the "multipotentiality" of the parts of the egg cell, (*i.e.,* their ability to assume a variety of roles, as needed) but also by their harmonious interaction. Integrated action is not found exclusively in living things. Mutually adapted parts and purposive organization are common also to purely mechanical, physical systems, such as machines. These are, in fact, the best example of what Driesch calls "static teleology" or "teleology of constellation." The structure as a whole has a purpose, and each part within it is designed to serve a particular function instrumental to the realization of that purpose. However, the goal and the adjustment to it of the parts are determined not by an inner agency, but by the human intellect which prearranged the integrated structure of the parts.

In the case of organisms, however, the parts are integrated in a more dynamic fashion. If one tampers with the normal functioning of a single part, this interference will be compensated by an adjustment of other parts. If you remove a part of a machine, it may continue to function with decreased efficiency, but no adjacent part will assume the task of the debilitated part. If you knock out the headlight of a car, the car will continue to run, but there will be illumination on one side only, and not even stronger illumination on that side. The missing part will not replace itself. But if you remove a portion of the egg cell in an organism such as the sea-urchin embryos with which Driesch experimented, the function of that portion will be assumed by some other portion which normally plays a different part. Now, Driesch asks, how could the parts "know" what compensation was demanded of them? This cannot be predetermined, for it is not possible to anticipate what accidents may occur. At best the organism can be equipped for all possible situations, but when the occasion arises, it must still "decide" which of its various solutions is called for. How is this selection made?

Driesch claimed that there must be some agent which determines which function is damaged and which potentiality of the organism will be released in order to compensate for the damage. This is the job he assigns to the entelechy. Thus the entelechy does not introduce any new features into the system, nor is it an extra vital force. It is not an Aristotelian guiding program which determines the realization of the organism according to its own species. It is, rather, a kind of governor, or corrective feature, which intervenes when normal development is threatened.

It is, however, also involved in normal development. As the organism develops, it increases in complexity. It does this not exclusively by the acquisition of new parts (which can be accounted for in terms of known physical principles), but also by an increase in the variety of relationships among the previously existing parts. What takes place is a transformation of a collection of parts, a sheer aggregate, into an organized and integrated whole. It is as if a heap of nuts and bolts and wires and pieces of metal, cloth, and insulation were suddenly to assemble itself into a car. But we know that this cannot occur without a guiding cause. In the case of the car we recognize the cause to be human craftsmanship; in the organism it is an internal principle, the entelechy.

Driesch refers to the entelechy as a "unifying causality," a cause which is neither external to the objects upon which it operates nor is one of them. Without introducing anything new into the set of those objects, it unifies them into a whole. It is this whole-making feature which is fundamental to the notion of entelechy. Driesch regards wholeness as a property of a system over and above the properties of its individual parts. Furthermore, if all the properties of the system require an explanation, then an explanation of the wholeness of the composite entity must be provided too. But this explanation cannot be identical with the explanation given for any of the constituent parts.

Just as Bergson invokes the vital force, so Driesch introduces the entelechy to fill a gap in empirical evidence: "Whenever what is known of other elemental facts is proved to be unable to explain the facts in a new field of investigation, something new and elemental must always be introduced." (H. Driesch, *Science and Philosophy of the Organism,* I, 142.)

Like Bergson, Driesch fills the gap with a nonempirical element. The entelechy occupies neither space nor time; nor is it a psychic or mental entity. It has no mass and is not a form of substance or of energy. It cannot be measured or quantified in any way. There cannot be more or less of it. It does not depend upon matter, but it is upon matter that it produces effects. And the effects that it has are not to increase the quantity of matter present in the world, but to suspend or to release tendencies which are inherently present in material systems.

Having thus described the entelechy, Driesch believes himself free of the criticism which is often leveled against proponents of the vital-force theory; namely, that the operation of the vital principle would involve an expenditure of energy in the universe for which there is no evidence. Driesch thinks that if you merely rearrange the existing parts of something, without making any substantial additions, then you do not use up any energy. But if you have ever cleaned up a disorderly room (adding nothing and subtracting nothing), you will know that this too requires energy.

A further limitation of Driesch's concept of entelechy as the life principle is that the entelechy can explain only how or why a particular organism realizes its particular destiny but not why it has this destiny. A lobster, for example, may manage to mature to lobsterhood despite this or that circumstantial impediment, thanks to the intervention of its entelechy. But why lobsters in general have the properties they do have, or how lobsters came to develop out of a more primitive species of organism, cannot be accounted for in terms of the entelechy which governs the growth of the individual lobster.

Driesch is aware of this restriction on his theory. But he suggests that there is perhaps a *suprapersonal* entelechy which governs the progressive complication of species and which in fact imposes a determinate order upon the whole of world history. Again, as in the case of Bergson's *élan vital,* one can offer no conclusive evidence against such a universal principle, but there is no evidence in its favor. And it is unlikely that anyone would be inclined to believe in it without a prior commitment. Once having accepted entelechies at the level of the individual, one may be prepared to do so again at the species level; but if one is initially unconvinced by the first entelechy, it is unlikely that he will become persuaded by the addition of a second and more abstract one. The reasoning here parallels that of our earlier instance in which the acceptance of one natural discontinuity smoothed the way to the admission of others. (See the discussion of theism and vitalism.)

In effect, the introduction of the cosmic entelechy weakens Driesch's theory. Far from being derived from empirical evidence, the doctrine of entelechies presupposes a belief in an orderly universe, governed by an agency which operates according to a preestablished harmony. Furthermore, the doctrine of entelechies does not explain how life begins and how it ends. By what means does the entelechy act on matter, and once united, why do entelechy and matter ever part company? Driesch himself denies the significance of such questions as compared to that of "the laws of life"; but surely any account of the nature of life should at least open some avenue of approach to the questions of life's beginning and end.

Driesch's doctrine of entelechies is generally held to be historically significant. To many people Driesch represents the end of the era of the scientist-vitalist. There still are scientists who are vitalists, but few of them have the boldness of Driesch. Many are scientists with one hand and vitalists with the other. But there are no longer any reputable adherents to Driesch's view.

Recently some Catholic scholars have updated Aristotle's entelechy, shifting its emphasis from the organism as a whole to the cell as the unit of life and attributing to it the power to "give direction to a process." The entelechy is held responsible for cell division and for the unique unity of the cell. The subcellular constituents, or molecules, are also organized composites of elements, but they are not alive and the entelechy is not held responsible for them. While a molecule consists of the same chemical elements as the cell, and may even be identical with the cell, it does not require an entelechy to account for its wholeness. It is not clear why this difference exists. However, the explanation offered is that the structure of a molecule is strictly a function of the interrelations of its components, while the structure of the cell is at least partly conditioned by the role it plays within the whole of which it is a part.

But this larger whole also requires an entelechy as the dominant unity governing the distinct unities of the separate cells. One may think of this arrangement as analogous to that of a federation of states in which each is a separate political unit, autonomous, and yet coordinated under a superstructure of law.

Once such a hierarchy of systems is admitted, however, there is no end to the sequence of systems. Cells are parts of organisms, which in turn are parts of biological systems, which in turn belong to ecological and social communities. These may be considered a physical system, and ultimately we are returned to the cosmic entelechy, a principle of order of universal scope. This, once again, is readily identified with God or an absolute spirit.

The entelechy conceived on these neo-Drieschian views may be regarded as the form of the system, inseparable from but not identical with matter. The material organization is an expression of the governing entelechy of the cell. It cannot be identical with it, for then the cell could not fail to achieve perfect self-realization. But without the entelechy, there could be no achievement even of imperfect realization.

The doctrine of entelechies hinges upon two central concepts, each of which has formed the basis of another vitalistic theory less obscure than the doctrine of entelechies itself.

On the one hand, the doctrine of entelechies rests on the prem-

ise that *organization entails novelty,* that a living being possesses a
new feature which is not to be found in any of its constituent
parts, but comes into being with their conjunction. One may pursue the source of this novelty and try to explain exactly what it is
that comes into existence. Vitalistic theories which stress the occurrence of novelty maintain that life appeared in the universe as
a kind of evolutionary emergent or new level of reality.

The other premise upon which the doctrine of entelechies rests
is the claim that the organization of living beings involves a
wholeness greater than the sum of its parts. This wholeness is
claimed to be self-sustaining, the organism having the tendency to
maintain its identity against adverse environmental conditions.
The claim is made that the essential feature of life is a kind of
self-stabilizing integrity that preserves its own wholeness. It is neither imposed by external forces nor is it necessarily newly emergent. Above all, it cannot be explained in purely physical and
chemical terms, for it differs from purely material organizations.

These two propositions, that life is a newly emergent phenomenon, and that life is essentially a self-maintaining organization,
represent the basis of two additional vitalistic theories. The first is
the doctrine of *emergence,* which will be discussed in the following section; the second is the doctrine of *organicism,* which will be
considered in the section after that.

Emergent Evolution and Emergence

We have seen that the doctrine of entelechies may be criticized
as failing to account for the beginning and end of life. It treats life
as if it had no history. If anything, it provides an explanation of
the preservation of the status quo. Assuming that we have a variety of species, the doctrine of entelechies offers an explanation of
how each species manages to reaffirm and maintain its identity by
self-replication even if anything should threaten it. But how did
the species begin, and why are there so many kinds of organisms,

each with a distinct identity? What brought about the first entelechy?

The vitalistic theory we are about to examine takes up those historical questions. In addition, it challenges a logical claim which is implicit in the doctrine of entelechies. In order to clarify our discussion, we will try to keep the historical and logical issues separate. But the distinction is not invariably made within the theory of emergent evolution.

Simply stated, the causal principle underlying the doctrine of entelechies is that you cannot get something out of nothing. The nonexistent cannot give rise to an existent. That which causes an effect must in some fashion "include" or "contain" the effect to begin with; otherwise, the effect would arise out of nothing. This principle has been defended in the interest of religious beliefs and it is consistent with common sense. Think of the folk tale of the goose that laid the golden eggs. The farmer who killed the goose in his greed expected to find a whole warehouse of eggs inside. It is not far from such popular beliefs to the preformationist theory, which was widely held by scientists prior to Darwin. According to this theory, the original seed or cell of an organism contained within itself a minute, but complete adult version of itself. In every human sperm a miniature but fully formed human body (homunculus) lived. This had only to grow and develop those faculties which it already possessed. Presumably it contained another still smaller homunculus, which in turn sheltered another. Ultimately, the whole race of men could spring from an original human who contained homunculi. Scientists who believed in preformation were called evolutionists. (Literally, the term "evolution" means to unfold or turn out.) They opposed the epigenicists who believed that the germ cell acquires new characteristics, one after the other, and does not simply unfold preexistent qualities. Epigenicists (sometimes called emergentists) believed that the human embryo, for example, gradually develops new features, first the neck, then the arms and legs, then the facial features, and so on.

The pre-Darwinian evolutionists, like the proponents of the doctrine of entelechies, held to the causal belief that the effect must be "contained in" the cause. This is why the mature features were thought to have "unfolded" or been "unpacked" from their germinal state. For the earlier stage as cause must be richer than the effect, appearances to the contrary notwithstanding.

It follows from this conception of causality that a full understanding of the cause would necessarily include an understanding of the effect, and hence the effect may be predicted with certainty. From the point of view of a perfect understanding, the entire future of a seed is included in its beginnings.

This concept of causality is implied by the classic doctrine of entelechies. Aristotle's entelechy is logically equivalent to the homunculus. It is the plan or program according to which the organism develops. Driesch's entelechy is somewhat different. Since the program of development is built into the material structure of the organism, the entelechy is required only for the complementary task of triggering corrective action where and when needed. Driesch's doctrine leaves room for somewhat greater indeterminacy than the classic view allows. But both positions, since they involve a foreordained program, really deny that there is novelty in the universe. If one thinks of the world as the intended product of a first cause, then, in principle, it should be possible to foresee the entire future development of the world from an understanding of that cause or of its intentions, just as the features of the mature man are held to be predictable from knowledge of the homunculus. Most traditional Western theologies attribute such foreknowledge to God.

But human beings do not have infallible foreknowledge. We are conscious of novelty in the world, and its presence is particularly striking among living things. As we have noted before, it is this very factor of indeterminacy and unpredictability which some people have designated as the distinguishing feature of the living as opposed to the nonliving. (See the discussion of Bergson.) Is there some way of reconciling a traditional view of causality with our conflicting experience of novelty?

The doctrine of emergent evolution is an attempt to bring about such a compromise. It affirms that the historically later stages of the development of an organism arise out of, but are not strictly deducible from, knowledge of earlier states. Evolution is the emergence of something new, something which cannot be predicted from earlier stages and is to that extent richer than its cause. The doctrine of evolution is not itself a causal theory. On the contrary, it presupposes a causal theory. To say that *a* is the cause of *b* is not to explain what causality is. This statement implies that the concept of causality is already understood. To say that *x* has evolved from *y* is to assume the prior acceptance of a causal theory. The post-Darwinian theory of evolution allows that the effect might be "greater than" the cause. In this modern sense of evolution, the effect is not merely an unfurling or display of features antecedently "rolled up" in the cause. It is wholly unprecedented.

Darwin was aware of the radical nature of this causal claim. He says:

> Nothing at first can appear more difficult to believe than that the more complex organs and instincts have been perfected, not by means superior to, though analogous with, human reason, but by the accumulation of innumerable slight variations. . . . Nevertheless, this difficulty, though appearing to our imagination insuperably great, cannot be considered real. . . . (*The Origin of Species:* Chap. XV, Recapitulation.)

He then goes on to explain the mechanism of natural selection, according to which chance variations produce effects, some of which are enrichments with respect to their causal antecedents. Darwin's theory has nothing to do with the origin of life itself, only with the increasing complexity of the various species of organisms. The doctrine of emergent evolution extends the pattern of Darwin's reasoning to account for the initial occurrence of life as well as for other outstanding events in the history of the universe, such as the appearance of consciousness and even of matter itself. These events are believed to succeed one another in time; *i.e.,* at some historical moment matter gave rise to life, which in turn gave rise to consciousness, which in turn yielded a product of still greater complexity.

Later versions of the doctrine maintained that successive increases of complexity were due not exclusively to internal factors of the organism, but also in part to its external environment. In effect, this modification of the theory reduces the claim to novelty in the universe. A particular complex effect is held to be the consequence not of a single simple cause, but of many causes combined. It is, however, implied that the result is unpredictable from an analysis of the component causes if each is considered separately. Furthermore, the effect itself is such that it is not merely a summation of the causes which produced it. We cannot predict the effect from the cause. We also cannot discover the cause by breaking the effect down into constituent causal elements. The effect is something new in the universe and may in fact be quite simple.

A combination is not simple if all its individual parts continue to function in an independent fashion when they are combined. This would be a complex phenomenon, whose action is the sum of all the individual actions of its parts. A combination is complex if one can differentiate many contributive actions within it. A simple entity or action cannot be so analyzed although its action may be dependent upon a complex interaction of parts and prior actions. These, however, are no longer evident in the action which finally takes place. The action assumes a character of its own which must be described in its own terms, and for this purpose a new vocabulary is required. We need a new *vocabulary* to say something about a situation, when we find that the result of a combination of entities cannot be fully described in familiar terms simply by enumerating the parts we have combined. Something happens when the parts are combined, such that the universe with the combination is richer than it would be with all the parts uncombined. We find that we must add to our stock of terms in order to describe the situation after the combination has occurred. It looks as if something new has come into being (emerged) and needs a name.

Imagine the tempting aroma and sight of a cake, freshly baked, standing on the kitchen table. Now think of that same kitchen table one hour earlier with all the ingredients unmixed. Sniff the flour, the eggs, the butter, the sugar. You will neither see nor smell the cake. The cake has a quality of its own which is not only distinct from that of its constituent parts, but which eclipses them. Once the cake is baked, you no longer perceive the individual features of eggs, flour, and butter. Their identity has been sacrificed to that of the cake. The cake is something new, simple, and emergent, and we call it "cake" to differentiate its level of existence from that of a mere complex of flour-butter-eggs-sugar and so forth.

The problem of the cake illustrates the dilemma of the philosophically inclined scientist. He knows very well what he put into that cake, and that no mysterious gremlins produced additional flavor and aroma. Yet there they are. The cake *has* flavor; it *has* an aroma. Where did they come from? Are they "real" in the same sense in which the flour and eggs which we know to be "in" the cake are real? Have we a metaphysical problem—the apparent coming into being of an effect which is richer than its cause? Or is the question purely logical? If it is purely logical, then it is not strictly correct to speak of introducing a new level of being or reality, but only of analysis. The baking of the cake does not produce a new thing or substance. However, it reveals that our earlier conception of what would happen if certain substances are combined was inadequate. We need more comprehensive physical laws. What comes into existence when you bake a cake is not a new entity, but a new application of a set of laws.

You may feel uncomfortable with this suggestion; for surely one does not eat a set of laws, and the cake is edible. But do not lose sight of the issue. We are concerned with the novelty which we experience or of which we can speak when the ingredients of the cake are combined. You could eat the eggs and flour and so forth separately, and, in principle, get as much nutritive value out of them as you do out of the cake. You do not add vitamins or proteins by baking a cake. You do not add edibility. But some things may be *said* of the cake which may not be said of its ingredients either separately or together. If we concentrate on the new things that may be said about the cake and avoid inferring that the cake must be a new thing, then we are substituting logic for metaphysics.

The person eating the cake is not concerned whether we think of his cake logically or metaphysically. Eating the cake would make it disappear substantially, but it would not affect its logical properties. Correspondingly, to declare that calling it a cake only confers a new logical distinction upon the butter, milk, and eggs does not affect their flavor. The metaphysical and the logical are two different ways of considering the same phenomenon. They deal with the same roughly delineated events of the world or of experience, but they do not coincide exactly. Nor are they mutually exclusive. We do not have to choose between them in the sense that if one is right, the other must be wrong. We might find that for some purposes one approach is preferable to the other. This is just one of many instances in life where we cannot say categorically that explanation *a* is right and explanation *b* is wrong. However, we can specify the reasons for preferring explanation *a* under the circumstances at hand, and the reasons for rejecting *b*. And we can indicate, generally, the kind of situation which calls for one type of explanation in preference to another. In doing this we may, simultaneously, clear up some of the historical disagreements and agreements which have been clouded by a failure to differentiate between different types of explanation.

The doctrine of emergence, or emergents, has suffered from such confusion. It is, however, possible to see a historical sequence of emergence theories which reflects a gradual shift from the metaphysical to the logical mode of explanation. The doctrine

of emergence was in its heyday during the latter half of the nineteenth century and the beginning of the twentieth. Proponents of
this view believed that new entities, metaphysical emergents, *came
into existence* in violation of traditional causal principles. They
were supposed to arise out of conditions inherently simpler than
themselves—not out of preexisting conditions which implicitly
contained them. Thus living matter was held to have evolved from
nonliving matter, and consciousness from the living in a continuous sequence that introduced something novel and unpredictable
into the universe. This view abandons the old conception of evolution as an unwinding of a process which is already implicit in its
initial stages. Evolution is now conceived in the contemporary
sense which involves a genuine production of novelty. Samuel Alexander (1859–1938), an English philosopher, and one of the
foremost defenders of the doctrine of emergent evolution, argued
that evolution as described by Darwin could be extended beyond
the species of living organisms to apply also to the development of
suborganic events and supraorganic events.

Alexander thought of the entire universe as having undergone a
historical process of evolution. New orders of being arose from an
elementary space-time condition, acquiring new complexities and
new behavioral regularities, which could be expressed by newly
formulated laws.

As we have seen in the case of the cake, however, the actual
locus of the novelty is not entirely clear. For, as in the case of the
cake, the new entities supervene upon the prior existence of the
ingredients with no new elements in addition to those ingredients.
How, then, does novelty arise? Lloyd Morgan, one of Alexander's
chief disciples, and a noted biologist in his own right, proposed
that emergent evolution gives rise not to new entities, but to new
forms of relatedness. Thus he promotes a shift to a quasi-logical
rather than strictly metaphysical account of emergence—from
things that exist to relations among things that exist. However,
since he regards this new relatedness as being effective as an
agent, *i.e.,* as responsible for a change in "the existing go of

events," it is apparent that he has not moved very far from the notion of a newly created metaphysical entity.

Since the doctrine of emergent evolution addresses itself to the historical problem of the origin of life, it is preferable in this respect to the doctrine of entelechies. However, it falls short of the Drieschian doctrine of entelechies in describing the nature of life and in characterizing the features which differentiate living things from inanimate objects. In effect, Alexander and Morgan merely reaffirm that life is different, that it is new, and that this novelty is superimposed upon the physicochemical complex which it presupposes. Their very emphasis upon the novelty of the emergent tends to obscure its nature; for if it is something radically new, then it must be indescribable in conventional terms.

This is not to say that we are condemned forever to ignorance of the nature of emergents. Once they have occurred, they may be described. Once we are acquainted with the conditions under which they have arisen, we may predict with some probability that they will continue to arise under those conditions. But this does not explain them; it only calls attention to the regularity of their occurrence.

This may however be a useful enterprise. Consider an analogous situation: suppose that you discover that every time you eat strawberries, you break out in a horrible itchy rash. Now there is no logical connection between strawberries and rashes, and for most people there is no connection at all. But for you, the observed regularity is very significant, for it enables you to make the prediction: my eating strawberries leads to my having rashes, and permits me to adjust my life accordingly. It does not explain the rash. Why should you have the rash when other people do not? Why from strawberries rather than from potatoes? If you are curious, you can go on to make further inquiries about the chemistry of strawberries and about your own physiology, and maybe you will find that you are sensitive to something in the strawberries. This would be more useful than just knowing about strawberries, because it permits you to avoid other things which contain that substance without having to go through the painful experience of getting the rash first. It might enable you to find an antidote which would counteract the substance in the strawberries. With such expanded knowledge, you could make bigger and better predictions than you could simply from the experienced association of strawberries and the rash.

The emergentist believes that we can predict the emergent from its antecedent (preexisting) conditions only by using a process of association similar to the one which led you to predict the rash from strawberries. Just as knowledge of the strawberries without experience of their association with the rash would not be sufficient to make the prediction of the rash, so, according to the emergentist, the occurrence of life could not be anticipated from mere acquaintance with the physicochemical conditions which give rise to it.

It is not altogether clear, however, why this unpredictability is present. In the case of the strawberries we searched for an adequate explanation—one which relates something *in* the strawberries with something *in* me. Such an answer to the question "Why do I break out in a rash after eating strawberries?" is more informative than "Because you have always done so in the past," because it permits confirmation of further predictions.

Do we have a parallel situation with respect to emergents? Assume that we know from experience that life always arises from specific physical conditions. Of course we do not know this. We do not have experience of the origin of life at all, and we cannot even be sure that it happened only once. We are certainly not in a position to say of life and its conditions, as we are of rashes and strawberries, that this is an association which we have frequently observed. But suppose we could make the association. What prevents us from going beyond that association to an explanation analogous to that which relates the substance in the strawberries to my physiology? Is this a matter of our ignorance or laziness, or some inherent unpredictability in the event itself? If there were something inherently unpredictable, how could we make the prediction in the first place? This is the issue which divides the emergentist from the nonemergentist. There is agreement that we now lack sufficient information to explain the appearance of life from nonliving conditions; but the nonemergentist says that this is because we have not studied hard enough and that the task can, in principle, be accomplished (*i.e.,* it is analogous to the strawberries-rash case). The emergentist declares that, whatever knowledge we may acquire of living things, and however we expand our knowledge of physics and chemistry, we will never be able to fully explain the occurrence of life in terms of physics and chemistry. This is because there is something radically new and inexplicable when life appears. Hard work will not yield the knowledge we seek, because some features of the world are simply beyond our understanding.

The emergentist claims that life is fundamentally distinct from the purely physicochemical. This does not preclude the application of physicochemical laws to living things. On the contrary, emergentists often distinguish themselves from traditional vitalists on the ground that they, unlike vitalists, believe that life is lawful. Emergentists claim that life is not exhausted by physicochemical laws. Their point is that the laws which apply to life are distinct from, but not necessarily contradictory to, the laws of physics and chemistry. At a certain level of organization, the latter laws are no longer sufficient to account for observable phenomena.

The emphasis upon distinct biological *laws,* as opposed to distinct biological entities or qualities, distinguishes explanatory (or logical) emergence from metaphysical emergence. Modern emergentists tend to avoid the issue of whether or not a new thing, such as the cake of our previous example, actually comes into existence when certain conditions are filled. They argue instead that we are justified only in noting that at a certain level of complexity of combination of physicochemical substances, new patterns of behavior may be observed. They agree with metaphysical emergentists that such innovations may occur at any level of complexity, not only at the stage where the inanimate gives rise to life. They employ such illustrations as the formation of water out of the combination of hydrogen and oxygen, or of salt out of sodium and chlorine, in order to indicate how common an event emergence is. The wetness of water, its qualities of surface tension, and its characteristics of dousing flames or quenching thirst are unique to itself and could not be discovered from an examination of its component parts. The water is not a substance over and above the oxygen and hydrogen of which it is composed. But when oxygen and hydrogen are combined in a prescribed ratio and under certain conditions, then a new set of descriptive principles becomes applicable to that combination.

The question may be asked whether the new properties could have been predicted had we possessed enough knowledge of the properties of the elementary particles? Or, alternatively, do the laws of nature consist in definite unabridgeable sets, which may be completely understandable within their own bounds, but which do not form a continuous series of logically related laws?

The logical emergentist, although he may withhold judgment about the actual coming into being of new entities, nonetheless is making a claim which is in its own way more far reaching than that made by the metaphysical emergentist. Most people who defend explanatory emergence believe in metaphysical emergence and are trying to justify their belief logically. It is possible to believe in the emergence of group properties without offering new explanatory principles. One would expect, however, that the positing of new laws follows upon, rather than precedes, the discovery of new entities. New entities suggest, but do not necessitate new laws, and new laws do not presuppose new entities. Some combinations of known constituents might yet be inexplicable in terms of any known law. Consequently, new laws may be required without any corresponding new substance. In this sense, explanatory emergence is the more fundamental of the two positions because it is presupposed by simple metaphysical emergence and is itself logically independent.

Both positions are compatible with, but do not require, histori-

cal evolutionism. Alexander and Morgan both believed that a sequence of emergences took place, beginning with the appearance
of matter from a primitive stuff and culminating in the emergence
of a deity. Explanatory emergence may also combine with a
chronological sequence of novel regularities. Belief in historical
order is not entailed by belief in emergence as such. The doctrine
of emergence asserts only that novelty does occur—that a given
set of circumstances which is fully describable in terms of its components and governing principles may nonetheless yield a situation
which could not have been predicted simply on the basis of prior
knowledge of those components and principles. Such novelty
might occur on random occasions.

This doctrine is far broader in scope than the issue separating
mechanism and vitalism, but it touches upon that controversy insofar as it is claimed that life is a quality which emerges from
purely physicochemical compounds, or that the laws of biology
arise from, but are independent of, the laws of physics and chemistry.

The crucial element of emergence theories is their insistence
upon the occurrence of novelty and their corresponding denial of
the continuity of nature. However, if one presses explanatory
emergence, one may end by minimizing the claim of novelty. Consider what is involved in the declaration that new sets of laws
arise.

A law does not exist in a vacuum. Laws refer both to theories
and to states of affairs which they are claimed to describe and explain. When a new law is "discovered" or stated, there is some
reason for it. Existing laws may inadequately account for the
known state of affairs. They may also require reformulation when
the theoretical framework, in terms of which they are stated, is
modified. In either case, it is not just the proposed law which is
new, but the entire context from which that law emerges. The in-
novation is not the law, but the set of conditions which call for the
law. This may mean that explanatory emergence is impossible

without metaphysical emergence. But it may also mean that the entire preoccupation with novelty is misplaced.

As we have seen, the doctrine of emergence has been associated with evolution and consequently with the idea that a historical progression of innovations takes place. An emergent is something new in time, if not in space. But consider how we ordinarily use the expression "to emerge" or "emergent." "She emerged, still dripping, out of the steaming bath." "The snowy peak of Mount Fuji emerged out of the mist." "There, to our horror, an iceberg emerged before the prow." "At last, his broadly smiling face emerged from the crowd." In all of these usages the term refers not to something new coming into existence, but to something becoming detached from or distinguishable from an environment. She got out of the bathtub. Mount Fuji became visible through the clouds. The iceberg was sticking out of the water. My friend's face was recognizable among the other faces in the crowd. In each case the object in question was *there* prior to being attended to as a distinct entity. There is not even any question of its being brought forth by or produced by the surrounding substance.

The emergence of the object is either the consequence of its own efforts (as in the case of the girl in the bathtub) or of a discriminatory act of attention on the part of the observer (as when I recognize my friend in the crowd). It may be due to the agency of, or the removal of, some intervening factor. The emergent itself was there all the time, but its relation to its environment is altered either in fact or in someone's estimation. The novelty that is introduced has no absolute status, but is entirely relative to the circumstances. Is there some "emergent" that remains when the girl has climbed out of the tub and dried herself? When the clouds have cleared, does Fujiyama still emerge? Does my friend's face continue to emerge once we have shaken hands and greeted each other? Surely not. To speak of emergence in these contexts is simply to call attention to particular, momentary organizations of experience. They have no fixed historical place, no prescribed position in either time or space, although they are neither atemporal nor aspatial. Above all, they do not in any meaningful sense constitute an introduction of novelty.

We might suggest that the fundamental commitment of emergence theories to novelty is really a historical accident. The real problem is not novelty, but organization of a certain kind. This leads us to the other possible outgrowth of the doctrine of entelechies referred to earlier—the doctrine of organicism.

Organicism

Like the doctrine of emergence, organicism stresses the notion of wholeness and organization. Organicists understand life to be the orderly integration of relations which hold among the parts of a complex system. Organization is regarded as a quality additional to the qualities which are organized as a whole, but not a quality on the same level as those qualities which are organized. It belongs to the whole in such a manner that the whole cannot be described as a mere assemblage of component parts. The whole is "more than" the sum of its parts; for it is those parts as a whole, or as organized.

Organicists tend to celebrate this quality of "wholeness" rather than to stress its novelty or unpredictability, but there are considerable variations among the views expressed by organicists. Some are very close to emergentist positions such as that of Lloyd Morgan, according to whom there is a new character which "emerges" when life appears; namely, a "new *relatedness*," or new manner of relating previously existent entities. Other organicists compare natural complexes or systems to one another, and note their systemic uniformity, maintaining that the differences between them are primarily matters of organizational complexity. Organicists, like emergentists, are not compelled to limit their claims about organizational levels of nature to the break between the living and the nonliving. This distinction marks but one of the many points at which a certain discontinuity is noted. At each level there is a gap between levels of complexity. It is, consequently, not the fact of organization which differentiates the living from the nonliving, for many things are organized, but rather the manner in which that organization takes place. There is some disagreement as to precisely what the particular nature of the vital level of organization might be.

Some organicists have concentrated upon the tenacity with which living things, while they are alive, appear to perpetuate their wholeness. They note a kind of dynamic equilibrium which living things maintain with respect to their environment. An organism does not simply offer resistance to its environment, remaining unaltered by external pressures; it adjusts to the environment. When you climb a mountain, where the oxygen is scarce, you also breathe more rapidly, taking in more frequent gasps of "thinner" air, and thereby inhaling more of the oxygen that is available. And when a bear hibernates in winter, he does not starve to death. He can afford not to eat for a period of weeks, because his entire metabolic process slows down so that the stored energy stretches until the spring. The stability that is achieved is a relational one. The organism is so integrated into its environment that in a sense it cannot really be distinguished from it. You are, after all, quite literally, what you eat, and also what you breathe and what you see and know.

There are other self-stabilizing entities besides organisms. A fire stabilizes itself as long as fuel and oxygen are available, and a thermostatically controlled central-heating system regulates itself to maintain a constant temperature. But, while self-stabilization cannot be regarded as uniquely identifying living things, nonetheless it can be taken as at least one possible indicator of life. Furthermore, the mode of stabilization which living things exhibit is rather remarkable. Instead of building purely protective walls about themselves, they interact openly and exchange parts with their environment, so that in fact they retain self-identity only by means of the continuity of their configuration. Neither their material constituents nor the pattern of their organization remains constant. Yet we can speak of an organism as a single entity from its embryonic stage through all the phases of its maturation. The question of personal identity has puzzled many people and poses fascinating philosophical as well as psychological, ethical, legal, and other problems, but it is not central to the issue of organicism as such.

Organicist biologists tend to concentrate upon certain features of organic systems which, although they may also be found in non-organic systems, nevertheless appear to warrant the attention of biologists because of their pervasiveness among organic systems. Foremost among these features is the one already referred to as "organic relatedness" or wholeness, sometimes expressed by the phrase, "the whole is greater than the sum of its parts." But the phrase does not clarify what the factor is by virtue of which the whole is greater than the sum of its parts.

According to some analysts the "greater than" element can only be expressed negatively. They point to the fact that a study of all the parts and of all the properties of all the parts, and of all the modes of action of all the parts and their properties taken in isolation *will not suffice* to explain the behavior of the whole. The action of the whole cannot be regarded as a summation of the actions of the individual parts, and, furthermore, even the actions of the individual parts functioning within the whole cannot be accounted for strictly within their own terms, for they are affected by their status within the whole. Just as a human being alone on a desert island would not behave in the same manner as in a city, so a single process, taken out of the context of the whole, will function differently from the way it will behave within the whole, for the wholeness of the whole is one of the factors which causally contributes to the behavior of the subordinate parts. Apart from the whole body of which they are a part, cells would grow and divide differently. This can be experimentally observed.

Imagine that we could know all there is to be said of the individual behavior and nature of each of the constituents of a composite whole. Our summation of these would not amount to a full explanation of that whole, and this would not be because of an oversight on our part or because we had somehow failed in our description of the parts. Rather, it is because each individual process is dependent upon its place within the whole. One cannot isolate a number of independent causal chains interacting within the whole, for the plan of the whole itself takes priority over these, and, so to speak, subverts them to its own ends.

However difficult it may be to describe, this is in fact a situation which is familiar to all of us. We all know how ideas are distorted when taken out of context, and how even inanimate objects, like furniture, are affected by the character of their surroundings. Certainly people, or rather the same person, vary a great deal when transplanted from situation to situation. In each case, the qualities of the unit are as much determined by its place within the whole as they are by its own internal constitution. The chair, its color and contour, is very much affected by the rug on which it stands, the curtains behind it, and its position within the room.

This is in no way an exclusively biological phenomenon. Some scientists have maintained that an organicist concept of wholeness is required to explain even the character of atoms and crystals. On the social level, it is a long-acknowledged observation that a comprehension of individual psychology is insufficient to account for mob behavior. The sheer fact of the surrounding crowd alters the behavior of its individual members. And yet in one sense the crowd is nothing more than a set of members. Nonetheless, it is alleged, crowd behavior is something over and above the combined behaviors of the participants and cannot be fully explained in terms of these. Persons who have been in mobs or participated in mass actions often admit that they engaged in activities which they would never have done alone. They were not simply imitating others or released from personal inhibitions by a sense of anonymity. They felt themselves to be a part of a larger whole, with a character of its own, one which included them, quite often swallowing up their own identity.

Likewise, if you look at the components of a picture, or listen to the parts of a melody, you will not succeed in reconstructing the work of art simply by conjoining its elements. The work of art is something independent of these. No mystical quality is required to explain what it is; indeed it is not just another element which is added to those already present. It is a quality of wholeness which is provided by the integration of the parts and which is itself instrumental in determining their integration.

One is not a vitalist just because he pays attention to biological "wholeness." Indeed, some people have argued just the reverse; that since organic wholeness is so widespread a phenomenon, occurring even at the level of particle physics, this is, if anything, an argument in favor of the ultimate reducibility of biology to physics. In other words it lays the groundwork for the denial of the autonomy of biology, and so undermines vitalism.

But more frequently organicism is a basis for vitalism, because it maintains that while the break between the living and the nonliving may not be the only discontinuity in nature, it is nevertheless a significant one. Some organicists try to analyze more closely the particular character of the "more than" quality.

An often-noted feature of organic wholes is their so-called *directiveness*. Organisms, both structurally and functionally, appear to be designed for a purpose. The parts are so organized as to suggest that there is a goal or end to which they are subordinate. Such goal-oriented organization is obviously not found exclusively in organisms, but it is a predominant feature of them. Furthermore, the presence of goal orientation is remarkable among living things in that it does not appear to arise from an external source. The goal appears to be self-determined and the organism has considerable latitude and adaptive flexibility in achieving the goal. In this respect organisms differ markedly, if only in a quantitative fashion, from mechanical complexes such as thermostats and computers, which may also be goal oriented, but whose organization is far more rigidly restricted. Mechanical organization depends upon the foresight of the organizer. The organization of organisms is held to be self-creative.

The notion of purposiveness or directiveness is not exhausted by pointing to the presence of a goal and to mechanisms for its achievement. When a stone rolls down a mountainside and stops at the bottom, we do not say that the action was purposive even though it involves passing through a sequence of steps from a point of initiation to a terminus. We can explain each phase of the motion of the stone in terms of the laws of gravitation, the slope of the mountain, the physical properties of the stone, atmospheric conditions, and other known factors. The action may be described as having a completeness or wholeness, insofar as it passes from an initial motion to a final state of rest, but there is no reason for attributing either an externally imposed or an internally ordained direction to the action, since it can be accounted for entirely in terms of the laws and initial conditions referred to.

But if a child rolls down a snow bank or a grassy mound, we have a very different situation. We can conceive of circumstances like those of the stone, where the child slips, loses his balance, and, like the stone, simply rolls down the hill until he lands at the bottom. But let us imagine him joyously letting go and with a shriek of delight "throwing himself" down the slope. The laws of gravitation and the atmospheric conditions remain unchanged. But something else in addition to these becomes pertinent. An element of "directiveness" is present, such that one cannot account for this occurrence without introducing references to this particular child and to his particular aims and purposes.

Many forms of organic behavior which once were thought to be literally and consciously purposive are no longer explained in such terms. The migrating behavior and nesting habits of birds, for example, and the spawning procedures of some fish are now accounted for in terms of temperature changes in the environment and correlated physiological changes in the organisms. It is not

necessary to attribute conscious purposes to these animals. We do
not know what a salmon is thinking, if indeed it is thinking at all,
as it swims upstream against the current to the place where its
eggs will be deposited. Since we have no basis for attributing conscious intentions to the fish, we should avoid drawing inferences
from our own experience. Nonetheless, the behavior of the fish appears, at least in part, to be governed by the end which is to be
accomplished, and not merely by antecedent circumstances. That
is, it is best understood by reference to the entire context of the
action, including its yet unrealized consequences, rather than by
simply enumerating prior conditions. The point here is that purposiveness may be attributed to an organic whole or totality which
could not be meaningfully attributed to an entity or process to
which the character of wholeness could not be ascribed. It is possible to refer to a process or part within a whole and to attribute
to it a function as a part which is determined by and consonant
with the goal orientation of the whole.

Such explanations make possible the accounting for deviations
from regularity as well as the regularity itself. It is one of the remarkable features of organisms that however repetitive their behavior may be, it is not always and without exception repetitive.
But the departures from order must be accounted for as much as
the adherence to order. It is profitable to an organism to reduce as
much as possible of its behavior to an orderly routine or habit.
But regularity can become mechanical beyond utility. We know in
the case of human behavior that habit, turned to compulsion or
obsession, becomes counterproductive. A mechanized or "fully
programmed" human being is one who has lost the capacity to
make reasonable judgment of priorities. Even a virtue, such as
cleanliness, if carried to excess, can lead to problems of health.
Constant washing, for example, in a dry climate, produces chafed
skin. The wise tourist learns to modify his behavior. A healthy organism retains the capacity to adapt even ritualized behavior to
the requirements of the end to be achieved. Sometimes immediate,
short-term aims must be sacrificed to the larger ends of the whole,

and the organism has the ability to make that adjustment. Some of the embryological variations described by Driesch and his followers can be explained in terms of the operation of such principles, without the introduction of a special governing entelechy. The subdivided sea-urchin eggs, for example, developed into individuals which were structurally complete, but they were reduced in size.

Organicists note a further characteristic of living organisms, differentiating them from other systems which may indeed be both organized and purposive. This is the internal origin of their purposiveness. A machine is designed by a human intellect and may be beautifully articulated in order to achieve the end it serves. But that end is imposed upon it. Its materials are selected with a view to the achievement of the end, and its parts are coordinated primarily with that aim. If the refrigerator is not functioning successfully, *i.e.,* in accordance with the purpose we assign to it, we examine it for a defective part, find it, and replace it with another which does enable it to run. The refrigerator itself is totally passive. It is reactive to the situation to which it is exposed (*e.g.,* it "works harder" when the external temperature is increased), but it does not go out and seek replacements for its own damaged parts or even make helpful suggestions. But a living organism bears its organizational principle as an active built-in character. Some philosophers and scientists have identified this indwelling formative character with the soul. In doing so, they come very close to the Aristotelian representation of the soul as the essence or form of the body, that which causes it to be the kind of thing it is. Thus as the axeness is to the axe—something which is imposed upon it by the will and intellect of the manufacturer—so in the case of the organism, its soul or formative principle is that which governs its being what it is from within. It is that which determines that the meat eaten by a dog becomes fur and paws of a dog, while the same meat eaten by a man would become hands and skin of a man.

While there is no doubt that external factors also condition the nature of a living thing, nonetheless there is a very considerable part of organic development which seems to be differently initiated. Some organicists account for this in terms of an indwelling principle, not reducible to any single part of any collection of parts, but operative throughout the whole organism. It is this principle which determines the nature of the organism, sets its aims, and orients it toward their realization.

Another feature of living organisms sometimes noted by organicists is their so-called *historicity*. Organic systems do not only have a history, which is true of many other things as well, but there is also an irreversible causal history of organisms. One aspect of historicity is the notion that organisms are the product of development. They take place in time—not merely in the sense that a sequence of steps is passed through—but in such a fashion that the particular character of each step is in part determined by the preceding step and could not have taken place without the prior occurrence of the preceding step. Thus the temporal order is fixed insofar as each stage presupposes its chronological predecessors and effects its own successors.

It is clear that a number of purely physical events can be explained as the outcome of historical processes. The erosion of mountains, the formation of river beds, even the distribution of gas molecules in a container presupposes the occurrence of a prior sequence of events. But organicist theory holds that living things are affected by different kinds of historical processes, or, at least, they are differently affected by processes which may in themselves be entirely physical. It is difficult to express just what these different types of processes or effects might be. The difference is sometimes elaborated as follows: In the case of a given physical event, say the emission by a substance of a radioactive particle, we may know that the emission is the consequence of a prior series of causes which we could specify. However, in order to describe the emission, we do not need to refer to those prior events. Furthermore, we can predict the occurrence of the emission simply from an acquaintance with the contemporary excited state of the substance. We do not even need to know the prior circumstances, even though they are causally relevant to the event in question. Some biological situations are analogous. But in some instances the very concepts that are employed include a tacit reference to historical factors. For example, to refer to an organism as a hybrid is not merely to describe its current genetic structure. It is not simply an abbreviated expression of the inventory of its genetic components, although it includes that. But more importantly the term refers to the parentage of the particular organism. It evokes specific historical factors. In psychology, such concepts as regression and fixation involve a similar reference to the specific historical past of the organism in question. The identification of a particular behavior pattern as regressive is not merely a description of it or a relation of it to present environmental stimuli. It is a way of relating this behavior pattern to a specific series of events in the individual past of the person. In other words, the historical reference is unavoidable.

We encounter such historical concepts in many areas of our experience. A veteran, for example, does not possess the property of veteranhood in such a fashion that you would find it by dissecting him, or by observing his behavior (although that might give you a clue). To call him a veteran is to say that included in his personal history is an involvement in a combat of a particular sort. Contrast this with *cripple:* Clearly a cripple is someone who has been injured or who suffers from a congenital defect. Yet, when we refer to someone as a cripple, we need only be aware of his present state. We can tell by looking at him, and we need not have any knowledge of what actually did take place in his past.

The organicists who argue for the independence of biology from physical sciences on the grounds of its historicity certainly do not mean to say that historical concepts are not applicable on levels other than the biological; nor do they say that all biological concepts are historical. An increasing number of so-called biological concepts are in fact turning out to be reducible to ahistorical physical concepts. The spawning behavior of the salmon, mentioned above, is a case in point. There is no need to refer to a prior history of the fish's desires. It is sufficient to describe the change of temperature in the stream and the accompanying physiological alterations which take place in the fish. Nonetheless, organicist biologists claim that there will always be a residue of biological concepts such as that of species, or embryo, which cannot under any circumstances be so reduced and whose elaboration necessarily involves some acquaintance with the past history of the whole individual or the species to which the concept applies. They regard this irreducibility as logically fundamental, possibly as based in the nature of things. Hence, the science of living things is guaranteed to be always and ultimately independent of the science of the nonliving.

There is one form of organicism which is based upon a somewhat different conception of wholeness than those which have been discussed so far. This is a doctrine propounded by students of *general systems theory,* which is a theory about science, or an approach to science, rather than being a particular science. General systems theorists are fascinated by the logic, or the pattern of reasoning which is employed in all the sciences, and which appears to provide a unifying bridge among them. In this logical unity they hope to find a correlation between sciences, which has not been discoverable on the supposition that there is a unity of subject matter or a metaphysical oneness to which all sciences apply.

General systems theory developed out of the realization that a certain analogy holds between all sets of events which are complex, having internal order, and which also exhibit orderly interrelations with other sets of events. Such interactive sets are known as systems. While their constituent elements may be widely disparate, they have a common feature in the order of their organization. Any set of phenomena which may be ordered as a system thus bears a certain likeness to any other set which may also be so ordered. The association between the parts of a system may be viewed as essentially communicative in nature. That is, wherever two or more elements are observed to change correlatively to one another, we may view them as related through an exchange of information. This is obviously plausible in the familiar systematic transformation of radio waves into intelligible sound patterns, but it is equally a reasonable interpretation of the thermostatic control of a heating system or the tendency of plants to turn toward the sun (heliotropism). Stressing the communicative element of systems relations, general systems theorists have made great advances in the comparatively new field of information theory, or cybernetics, which have had great practical consequences in the development of electronic computers and communications media.

Organicists are not primarily concerned with the practical success of general systems theory, but rather with the illumination it throws upon the part-whole relationship. General systems theorists are interested in discovering a system of systems. Their contention is that while distinct systems differ from one another in virtue of the particular integration of the constituents of one system as opposed to that of another, they resemble one another insofar as they are systems. They also believe that it is possible to approach a single set of correlated phenomena as a system, and then again to associate that entire system within a systematic superstructure. A set of events which comprises a system from one point of view may thus represent a constituent from the point of view of another system. Then system *a* would be a part of system *b,* but there would also be an analogical, as well as part-whole relation between them. This is complicated by the fact that a single event or phenomenon may occupy a place in radically different types of systems. A gadget on a car, for example, a radiator ornament or tail fin, is a part of the physical system of the automobile, produced and mounted on the assembly. But within the socioeconomic system of American society, those same gadgets have a nonmaterial prestige determining status. The role played is thus vastly different, and yet systems theorists hope to find analogous patterns even between systems as divergent as these. Ambitious systems analysts would like to see all things in the universe related to one another in terms of an elaborate network of systems.

According to this view, systems are themselves ordered into a system such that there is a hierarchical relation of inclusiveness between them. This does not mean anything quite so simple as the old children's toy (Chinese boxes) in which a box is contained in a box which is itself contained in a box and so on until one whole box is sufficient to contain the universe. Systems overlap with one another. They are not hermetically sealed off from one another. Nor are they found objectively in nature ready labeled and neatly stratified. It is rather the case that we discover them as reflections of our own explanatory concepts and principles. Within an organism we may concentrate upon the nervous system, the circulatory system, the lymphatic system, the reproductive system, or any other systematic relation among parts. In one sense these systems are all on a par. They can be represented on a chart relating the parts of the body. This level of organization is distinct from that of, say, the system of interpersonal-social relations or of atomic organization of protein molecules.

While the atomic structure of the protein molecule of, say, hemoglobin or of the nucleic acids can also be schematically represented, it would not be displayed on the same chart as that which exhibits the digestive system or the central nervous system. Similarly, one would not expect to find a restaurant replacing its regular menu listing sandwiches with a set of descriptive specifications of the protein and carbohydrate content of those sandwiches.

The question remains: might there not be an all-encompassing system which would permit the inclusion of these systems as well as the psychosocial and the political system within which those proteins and the digestive processes which consume them take place? Systems theorists as well as their opponents must ask whether or not the levels of system organization are so related to one another that one may be explained in terms of another. This is not an easy question to answer, for, since the organizational levels themselves are abstract representations of the concepts and principles which we find interesting or useful in explaining phenomena, so the affirmation of the reducibility or nonreducibility of one level to another must also reflect the theoretical convictions which we hold regarding nature.

Everyone will agree that there are some systems which may be wholly accounted for in terms of systems at another level, whether or not this is in fact a desirable thing to do. Thus the chemical properties of molecules can, at least in principle, be fully accounted for in terms of their subatomic structure. Everyone will also agree that for some purposes it is not useful to even seek level reductions. For the astronomer plotting planetary courses, it is not of much use to study the chemical constitution of the planets, and for the physiologist or the geneticist it may be more useful to concentrate upon the biological phenomena which confront him than to move to the level of physics. But this is not a categorical denial of the reducibility of the laws and concepts of biology to the laws and concepts of physics. It is a *heuristic* claim, an appeal to practicality. In effect it is an exhortation to act *as if* something were the case, not because we have any conclusive reason to think it is, but because it is fruitful with respect to a given goal we have set ourselves. That tells us nothing of what would be the case if our goals were to change or of what is the case independently of our goals.

General systems theorists are not necessarily organicists with respect to biology. Some choose to remain aloof from the question of reducibility. Some suggest that biological laws might well be reducible to physical laws and concepts, but regard this as a task so cumbersome as to be not worth the trouble. But many do believe that the levels of organization, at least at the point where the inanimate becomes the base of the animate, are inviolate. We may cite as an instance of this position the view of Ludwig von Bertalanffy, one of the founders of organicist biology and also an early adherent of general systems theory.

Bertalanffy affirms a mode of "biological uncertainty principle," according to which it may be possible to make statistical generalizations about "higher-order" *integrations,* or organizational levels; but it is impossible to deduce these directly from lower-order descriptions. Nor can lower-level truths be inferred from higher-order descriptions. The building up of higher levels of organization from the lower levels will always involve new laws, he says, which are not deducible from the laws of the lower levels. He concludes that the hierarchical mode of organization is of particular significance for the living organism and constitutes a fundamental principle of biological law. It is this irreducibility of levels to each other that guarantees the "autonomy" of the biological organism and of the science of biology. Biological concepts have their own mode of explanation, according to this thesis. They cannot be derived from the principles of physics.

Like the organicists, systems theorists lay great stress on organization as a fundamental biological principle. On the whole, they avoid animistic interpretations of organization as itself an active force or agent, having inscrutable magic power. They regard organization as a relation or a system of describing relationships— neither as subject nor as mode of action.

But because of their interest in organization as such, and in the establishment of a science of organizations as a supersystem of all sciences, general systems theorists are inclined to objectify systems as if they were the fixed crystalline spheres of Aristotle's astronomy. But astronomers now say that the stars and planets are not carried about in solid spheres; rather they rotate in orbits fixed by their own material structure and the attractive and repellent forces of things in their environment. Similarly, levels of explanation do not reflect only the qualities of the things which are to be explained but also the principles of explanation in terms of which we accommodate ourselves to the rest of the universe. These principles are likely to change from time to time as our interests and conceptual equipment change. We have no reason to deny that a given pattern of organization might be explained in terms of a more fundamental pattern. Assuming that physics itself were complete, there would be no reason to deny its sufficiency to account for biological phenomena. But there is also no reason to affirm that biology should be reducible to physics. Furthermore, this is not the sort of thing that can be discovered simply by looking or some other simple test. To a large extent it is a matter of decision, or, at least, it is dependent upon our selection of the concepts and principles in terms of which we choose to understand nature.

Toward the end of the nineteenth century, particularly among continental European intellectuals, it became fashionable to think that all sciences involving predominantly human factors, *e.g.,* history, anthropology, or psychology, required a mode of knowledge fundamentally dissimilar from that appropriate to the purely physical sciences. For centuries knowledge had been conceived as essentially uniform. Some distinctions had been drawn between theoretical and practical knowledge or between knowledge derived from rational insight and knowledge based upon revelation; but these were distinctions having to do with the sources or objects of knowledge. Now a radical distinction was drawn between actual modes of knowing: understanding (Verstehen), on the one hand, and knowing in the traditional sense (Wissen), on the other. It is not possible for us here to elaborate all the differences between the kinds of knowledge or to explore fully the reasons for believing that they exist as distinguished. The point of primary interest is that when one affirms that a category of things is approachable only by one form of understanding or mode of knowing, while another category is best studied by another, he has committed himself to a basic incommensurability between those two categories, so that it no longer makes sense to ask whether the two categories can be collapsed into one another. What I am suggesting is that the conclusion of the inquiry is implicit in the initial premise that two modes of investigation are called for. What should there be to call for distinct modes of knowing, if not some incommensurable quality of the objects to be known?

To say that two distinct methodologies are appropriate to two different objects is to say at the outset that the objects are distinct.

And if we then investigate them in accordance with the two pre-scribed manners, it is virtually impossible that we should come up with the conclusion that our objects are, after all, not radically distinct. The investigation is, in effect, a practical reinforcement of our initial conviction.

It was this dualistic mode of thinking about inquiry which stim-ulated much of the twentieth-century vitalism. The belief that liv-ing things are fundamentally different from nonliving ones is rarely the outcome of dispassionate investigation. It is generally adhered to prior to any investigation, and the investigation is meant to indicate what the nature of the difference is, rather than to demonstrate that there is a difference. Every one of the vitalis-tic views we have considered may be regarded as an attempt to ra-tionalize a phenomenon whose presence was independently af-firmed. Mechanistic attacks on vitalism are sometimes purely negative, aiming primarily to show that the suppositions of vital-ism are superfluous and unnecessary. Historically, mechanism has taken the offensive, systematically undermining the vitalistic argu-ments for vital forces, entelechies, and purpose in nature. In the face of all of these objections, vitalists have maintained that how-ever inadequate their own explanations may be, nonetheless there is a difference between the animate and the inanimate. This is evi-dent to everyone. Mechanists cannot explain it away, nor can they explain it. The burden of proof, according to vitalists, therefore falls not upon themselves, but upon the mechanists. For the mech-anist, too, starts from a conviction which he proceeds to justify by means of a process of reasoning which he regards as appropri-ate. Furthermore, the conviction of the mechanist, that the living and the nonliving are not fundamentally distinct, appears to vio-late all the tenets of good sense and of traditional education.

In the following chapter we will consider the pattern of mech-anistic reasoning. We shall follow its historical shifts to accom-modate changes in vitalistic thought, and we shall trace out the modifications demanded of it by the changing claims of its own commitment.

CHAPTER THREE

The Changing Face of Mechanism

THE various forms of vitalism, or rather instances of vitalistic theories which were discussed in the previous chapters, were not represented as following a historical sequence. They were distinguished because of logical or metaphysical differences between them, and historical instances were selected to illustrate the types of vitalistic theory. One might in fact hold a combined theory which adapts and selects features of several of the positions discussed. Nonetheless, there is a historical order which is not entirely accidental, for the positions fall along a continuum ranging from what might be called extreme vitalism to near-mechanism. As we have seen, the doctrine of organicism takes a variety of forms, and some of its proponents explicitly divorce themselves from the vitalism-mechanism controversy by declaring that both extremes are to be rejected, and that what they endorse is a third alternative.

There is much to be said in favor of such a third position, particularly insofar as the terms mechanism and vitalism have become so heavily laden with historical associations that it is difficult to use them without assuming an unnecessary emotional burden. Nevertheless there are also advantages to preserving the original polarity of concepts. Let me defend this semantic conservatism with a fanciful illustration.

Suppose that in some far-distant future, children are produced exclusively by artificial insemination. Assume further, that society exercises far-reaching control over the genetic components of the children which are produced. Genetic attributes are determinable at will such that specific physical features and traits of character can be selected. Imagine a special bank in which the carefully compounded and classified sperm is kept. Under appropriate circumstances women are impregnated with sperm which has been carefully selected to match their own genetic properties. The characteristics of the offspring can be predicted with a high degree of accuracy, and it then becomes possible to manufacture people to fit the needs and aims of the society as a whole. (This fantasy is not far from the fictional utopia of Aldous Huxley's *Brave New World,* but it is also entirely consistent with the projections for the future which geneticists are now making. It is tantalizing to discuss the moral and political implications of the proposal, but that would take us far beyond our purposes here.)

Let us now assume that after these children are born, they are raised in strict accordance with the plan of society. The state is fully responsible for their health, education, economic security, and so forth, but the children are actually cared for at least during early childhood by their natural mother and a male companion. The children may spend much of their time in public institutions, but they nonetheless have a family environment and a set of parents because it has been scientifically determined that there are certain emotional needs whose satisfaction is essential to the development of character and which can best be filled in the traditional family environment.

Now suppose that one day a group of theoretically inclined people decide to study the structure of society and to articulate the principles by which it operates. Professor *A* affirms that children are brought up by mother and father under the guidance of the state. Professor *B* objects violently to the use of the term "father." He contends that fathers no longer exist. Since the children are produced by artificial insemination, and since the sperm is provided by the bank which stores it acccording to genetic classification and not according to donor, there is no way of determining who the biological father is, and certainly no reason to assume that it is the husband of the mother. Furthermore, the husband has none of the traditional (nineteenth-century?) responsibilities of fatherhood. In fact he is functioning *in loco parentis* for the state. Perhaps the state itself comes closest to having those properties which once characterized the father. But since there are such obvious differences, it is best not to confuse the issue by identifying the state as the father. To do so would simply reflect a nostalgic tenacity favoring antiquated and outmoded institutions. Clearly what has happened is that society has evolved (progressed?) beyond the need for fatherhood. Fathers have become extinct, like dinosaurs.

Now along comes Professor *C,* who compliments his learned colleagues for the thoroughness of their inquiry, and offers the solution to the problem. What is needed, he says, is a third alternative. His colleagues, entrapped in a polarized conflict, are engaged in a battle of words. He proposes that the term "fatherhood" with all its historical and emotional associations be abandoned and that a new, clean, and scientific word, such as "proctorhood," be adopted to designate a new institution which preserves some of the elements of the old fatherhood and gives up others, but is really an independent concept which should be analyzed on its own terms. Once the historical debris is cleared away, we can settle down to an unbiased examination of the laws of the prevailing institution. The question, "Do children have fathers?" then becomes meaningless. It is neither true nor false, but simply empty of meaning.

What has happened here? Has any issue been resolved or have the issues merely been subtilized out of existence? Is there not sure to be a lingering dissatisfaction, a feeling that by verbal manipulations Professor *C* has shoved the real issues under the rug and turned to other matters, which, though equally meritorious of consideration, are nonetheless evasions of the issues posed by Professors *A* and *B?*

I am inclined to think that they are, and that a similar situation holds with respect to organicism. It is not a third alternative to mechanism and vitalism, but a shuffling confusion of the real question which divides them. Mechanists argue in favor of a continuity of nature such that the laws and concepts in terms of which the physical universe can be made intelligible are also sufficient to explain living phenomena. Vitalists deny this, maintaining instead that there is a break in continuity (at least one, and maybe more) such that additional laws and explanatory concepts must be introduced in order to account for life. Since these are contradictory positions, it is impossible to opt for a third alternative, and organicism does not do it. Some forms of organicism do come very close to a mechanistic affirmation of continuity. Either they declare that there is an analogy between the separate sets of laws applying to organizational levels, or they proclaim a logical superstructure, a system of systems, which supposedly unifies all laws and makes the discontinuities of nature appear less offensive. It seems to me however, that in the end organicism takes its place on the side of the vitalists.

Having placed the organicists in the camp of the vitalists, let me now defend them, from another point of view, for their refusal to ally themselves with the historical doctrines of vitalism. With respect to vitalism, their judgment was sound; in the case of mechanism it is based upon an anachronistic conception of what the doctrines of mechanism actually are. For mechanism, like vitalism, has undergone historical development. Organicists must concentrate their attack on the mechanistic doctrines which are contemporary—not those crude mechanical models which modern mechanists themselves would be (and have been) the first to reject.

The history of vitalism is reasonably well documented. It is clear that there has been a continuous retreat of vitalism before the encroachment of scientific explanations which are, for the most part, mechanistic. This explains why the arrangement of vitalistic theories in inverse order of their vitalistic intensity is also their historical order, although logically it might not have been. We may list a few of the major surrenders of vitalism.

Throughout the early period of Western Christendom it was believed that living phenomena fell into the domain of the divine and could be accounted for only in supernatural terms. Primary interest, of course, was in matters of human concern. Thus, events now regarded as "natural phenomena" were thought to have a moral or prophetic significance. A typical belief was the conviction that disease and madness were caused by visitations of supernatural creatures (demons). These were thought to act on physical substances, but neither they nor the actions which they performed were in any way supposed to be governed by, much less restricted to, physical laws. The notion of physical law itself was not yet stable.

In the seventeenth century, great advances were made in science which cast doubt on the independence of vital phenomena. It became philosophically respectable to believe that all of nature could be understood in terms of relatively few natural laws which were not interrupted, even by divine fiat. (God in his goodness would surely not do anything so confusing to mankind.) Conceptual tools and instruments of observation, such as analytic geometry and the microscope, were developed and lent support to the conviction of the universal application of natural law. More to the point, such investigations as those of William Harvey, which approached the circulation of the blood through the body as essentially a problem of hydraulic engineering, led to the view that animate bodies are not radically different in kind from inanimate ones, and are subject to the same laws.

Still it was believed that organisms may be differentiated from inorganic things in virtue of the presence of certain substances which could only be manufactured by living things. This belief also gave way to the advance of science when, in the nineteenth century, it was found that the so-called organic compounds could in fact be synthesized in a laboratory out of purely inorganic substances. One of the most famous instances of such an experiment was the one carried out in 1828 by Friedrich Wöhler. He obtained urea, a common organic compound which is manufactured in the kidneys of living creatures, by evaporating ammonium cyanate. As we have noted (p. 26), the term "organic" as used in chemistry was originally employed to refer to those substances which were thought to be obtainable exclusively from living organisms. That meaning is now obsolete, and although the term is still used, no one is quite sure what its modern connotation is. Sometimes it is defined as designating chemical compounds which contain carbon, but there are well-known carbon compounds, such as carbon dioxide, which are not classified as organic. This is only one of many instances where a historic error entraps us long beyond its correction by way of its influence on our language and, consequently, upon our thought.

As we have seen, vitalism did not die with the demise of the vital substance. One mode of its persistence was in the form of a belief in a vital energy. Once again, this doctrine was undermined by a scientific theory, the principle of the conservation of energy. The belief was long maintained that living things were exempt from this principle, and that their activities were governed by a special nonphysical form of energy. But as techniques of measurement became more precise and more biological processes were understood, it became evident that no special forces were required. The equilibrium of incoming energy sources and outgoing exertions could be accounted for in living organisms as well as in inanimate bodies.

This extension of the law of conservation of energy was made as recently as the beginning of the twentieth century, and since then the remaining vitalists have, with the exception of a few deviations, adhered to their views on the basis of predominately organicist (or holistic) arguments. Throughout the history of vitalism and most explicitly now (since vitalism has been slimmed down to its essentials) their fundamental conviction has been that biology as a science relates to phenomena of such a complex nature that the concepts and principles with which it deals are on a "higher" level than those of physics and chemistry, and cannot be reduced to them. In other words, biology is an independent science.

At present, the debate continues to agitate both scientists and philosophers, and it remains to be seen whether further scientific discoveries will have anything to do with the continued weakening of vitalism. On the whole the issue has become "sublimed" or reduced to its methodological essence, and it may well turn out that substantive scientific discoveries have no further impact on the controversy. This is a question which will require further discussion later.

Having briefly reflected upon the rather profound alterations which vitalism has undergone as determined by the context of the history of science, we ought now to recognize how unreasonable it would be not to expect a parallel history of alterations in the doctrines of mechanism. But in fact, very little attention has been given to this matter, and that is the anachronism to which reference was made above (p. 74). Modern vitalists disassociate themselves from historical vitalism because they recognize that it has been forced to give up many of its claims in the face of scientific investigation. They also disassociate themselves from mechanism, but they have not followed the history of mechanism, and they are under the misconception that it is the same today as it was in the seventeenth century. (It should be noted here that mechanists, too, are not free of such misconceptions. Frequently, in their attacks upon vitalism, they set up an antiquated version of it as their objective and proceed to knock it down with relative ease. But the vitalists themselves would be the first to applaud this operation and it falls far short of the purported aim—the disqualification of modern vitalism.)

The history of mechanism goes back to the dawn of Western philosophy. The earliest-known Greek philosophers sought to explain all of nature in terms of some single unifying principle such as fire or air. They believed that modifications of this primal stuff would be sufficient to account for all natural phenomena, including life, but they were not very clear on how and under what circumstances the modifications took place. It was also not clear how there could be a single unifying principle applicable to all things if in fact the things we experience are so varied and multitudinous.

Somewhat later, a group of philosophers called atomists declared that there was indeed a single kind of ultimate stuff, but that it was quantitatively plural. That is, there were many things, qualitatively alike, and all dispersed throughout a void, forming various temporary combinations, which accounted for the differences between things on the level of experience. One of the foremost of these philosophers was Democritus (460–362 B.C.), who believed that atoms are all alike, except in their shapes and sizes, and that they combine together as a result of collisions which take place when they are thrown about in a motion which is also natural to them. When they combine, they form patterns, some of which are more stable than others because of the shape of the atoms. The greater the stability of the pattern, the greater its likelihood of being repeated and persisting. In this fashion things of different kinds arise, and they possess different attributes, some of which are characteristic only of combinations, and not of individual atoms. Thus, from a single kind of material stuff it is possible to form all kinds of complex entities, including living ones, which, according to Democritus, differ from nonliving ones simply in being composed out of a larger proportion of tiny, round, and therefore swift and volatile atoms.

Democritus' theory is enchantingly simple, but it cannot explain many of the complex events and processes which take place in the inanimate physical world, to say nothing of the even more complicated activities of living organisms. Still, the fundamental principle which underlies Democritus' atomic theory has been professed by all subsequent forms of mechanism. This is the thesis that there is a continuity in nature such that all its phenomena can be comprehended in terms of universal law. Particular mechanistic doctrines have changed as our views of the nature of the universe itself have changed, but the fundamental commitment has remained constant, just as, in the case of vitalism, there has been a perpetual commitment to the thesis that life introduced a basic discontinuity.

The mechanistic theory that is most commonly taken as the prototype of mechanism is the view that prevailed in the seventeenth century. This is the view that was popularized by Descartes and elaborated to the point of caricature by his followers. Descartes was heir to the trend noted above of anti-supernaturalism and of reversion to Greek principles of the unity of nature. If a single set of mechanical principles could account for all terrestrial events, then why would the same principles not suffice as well to explain the living phenomena which occur within these systems? Descartes regarded the living organism as a kind of machine, powered by heat, which was generated in the "furnace" of the heart. This same heat, rarefied and moved by pressure, activated the brain and accounted for a great many of those experiences which we might now classify as purely psychological. Emotions, such as hate and love, and fear and joy, were all regarded simply as material perturbations. Beyond a doubt, organisms were extremely complex machines, and of them all man was the most complicated. But for all that, they were machines, entirely intelligible in terms of the mechanical laws of nature.

This is not to say that nothing whatever differentiates man from the rest of the universe, and from other animals in particular. Descartes was a mechanist insofar as his beliefs about life are concerned; but *thought* is something else. To explain it, Descartes maintained that there is a special thinking substance (the soul), whose chief activity is to think and whose essential nature is radically distinct from that of matter. It is in this belief that Descartes' more radical followers deviated from him, for they believed that even thought could be mechanically explained. One of his most outspoken disciples was Julien de la Mettrie, author of *A Natural History of the Soul* (1745) and *Man as Machine* (1747). La Mettrie argues that there is no evidence for the existence of an immaterial soul, and that the various activities which have been ascribed to it can be at least as well accounted for as functions of the material system.

Other followers of Descartes include a school of physiologists, known as the "iatromechanists," who applied Descartes' generalized theory of the mechanical activation of living organisms to specific physiological functions. They studied such organic activities as digestion, blood circulation, and respiration, and explained them on the model of mechanical processes. Thus, digestion, for example, was thought to be essentially a process of acid decomposition.

These scientists regarded the universe as a giant clockwork mechanism which operated in strict accordance with a limited number of intelligible laws. Any sort of motion could be explained in terms of the structure and subordinate motion of the parts of the object, and the motion of organisms was no exception to this rule. It is noteworthy that the only principle of change which was acknowledged at this time was motion in space. Matter was defined in terms of spatial extension, and any modification of matter was due to collisions of particles, to some form of pressure upon or inertia of parts, and to the consequent formation of new configurations.

While it was the principles of the mechanics of matter which set the tone of seventeenth-century mechanism, mechanism, even at this time, did not necessarily imply a commitment to materialism. It is true that La Mettrie was a thorough going materialist, but Descartes himself may serve as an example of a nonmaterialistic mechanist. Descartes believed that the life of organisms could be accounted for as a function of the heat of the heart, but he found it necessary to postulate a totally different substance—mind—to explain the process of human thought. Another of Descartes' successors, Gottfried Leibniz, far from affirming materialism, denied the existence of pure inert matter altogether. Like his twentieth-century disciple, Alfred North Whitehead, Leibniz believed that biology is the study of large organisms, while physics is the study of small ones. He was a mechanist in that he believed that the continuity of nature guaranteed the applicability of the same natural law on all levels of being; but he thought of all these levels as including living entities. The only difference between all these entities is that some are more conscious than others; and those that are most passive and inert are the ones we identify as nonliving matter.

It is also worth noting that the seventeenth-century mechanists were not atheists. While God was not absolutely required by their doctrine and a few considered the possibility of his nonexistence, most of them believed in a Creator who had made the initial clockwork and laid down its laws.

The mechanism of the seventeenth and early eighteenth centuries was dominated and limited by the character of contemporary science. The Newtonian laws deal with the motion of objects in space, and the machines constructed in Newton's time used these principles to translate the motion of one object into the spatial dislocation of another. This same machine image was reflected in the mechanistic conception of living phenomena. As we have seen, the physiology of the Cartesians and iatrophysicians was based upon a model of corpuscles moving at differential rates in space and exerting pressure upon one another, like the coordinated parts of a clock.

In the subsequent period of the late-eighteenth and early-nineteenth centuries, as the sciences of physics and chemistry developed, so did the mechanistic conception which was based upon it. Physiologists of the seventeenth century had regarded animal heat as the mechanical by-product of agitation, or friction. But in the eighteenth century, the living organism came to be regarded as a chemical heat transformer. Lavoisier drew an analogy between respiration and the burning of fuel: the animal body came to be represented as a combustion machine ingesting oxygen as well as food in a cycle of fuel-energy input, exercise, consumption, degradation of products, and elimination of chemically transformed waste material.

Mechanism continued to view the machine as the prototype of the organism, but new and different kinds of machines came into fashion. The laws of physical objects were no longer determined exclusively by additions of spatial motion, but also involved transformation of chemical substances and transference of energy. Correspondingly, the exemplar machine to which mechanists referred was no longer the clock or the steam engine with its transformation of fuel into energy. Well into the nineteenth century, the living organism was conceived as a kind of furnace, stoked with food and air which it transformed for the performance of its various functions. As the prototypical machine shifted, so did the interest of mechanists switch from certain characteristic functions and features of organisms, such as locomotion and anatomy, to other processes and features, such as general metabolism and the various physiological systems which maintain it.

Since the nineteenth century, a number of scientific advances have affected the character of modern mechanism. One is the fact that as the evidence has mounted up against a plurality of kinds of substances and/or forces to explain the differences between kinds of things not merely on the living, but on all levels, there has been increasing emphasis on factors of organization. Vitalists have turned this emphasis to their own ends by introducing a special organizational factor or directive principle to account for life; but mechanists note that organization is common throughout nature, and they seek to understand organizational differences between the living and the nonliving by analyzing the organizations themselves with great care, rather than by postulating external causes of the organization.

Vitalists sometimes point out that the tendency of living things to increased order is a violation of the second law of thermodynamics, according to which physical systems normally tend toward disorder or increased entropy. If you do not put energy into straightening up your room, it will eventually become a pile of junk, and this is the natural tendency of physical things. It is necessary to intervene to preserve their order. But living things, as long as they are alive, tend to preserve their order all by themselves. Their equilibrium is maintained by the processes of metabolism. If you put an organism on the top of a mountain where the oxygen is rare, it will breathe more rapidly than normal and thus inhale more frequent though smaller quantities of oxygen. If you put a human being in the heat where his normal body temperature would be likely to rise, he will perspire and the liquid will evaporate from the surface of his body, thereby cooling it and maintaining the normal body temperature.

These things, and many more like them, happen normally and automatically. The organism seems to function not merely like a very intricate machine, but like a self-regulating one. It is like the thermostatically controlled heating system of a house. When the temperature reaches a certain level, it triggers off a sensitive mechanism which turns off the furnace, and when the house has cooled off again to a predetermined temperature, the furnace turns itself back on again. But in the case of the heating system, we know who programmed it and how the electric circuits between the connections work. You can tamper with it yourself by changing the setting of the thermostat. There we have the organizational principle in the mind of the engineer who conceived it. But where is the plan in the case of the organism? Vitalists look for it in some alien principle analogous to the engineer (an entelechy or vital force). Mechanists seek it in the actual constitution of the physical entities which are organized.

This brings us to still another mechanical model, one which has dominated twentieth-century endeavors to understand living phenomena mechanistically. This is the electronic computer or communications machine which receives, stores, transmits, and utilizes information. Obviously, machines such as the nineteenth-century ones require an influx of energy which is transformed and used. The twentieth-century preoccupation is no longer with the source of power and its dispositions, but more specifically with the use which the system makes of that power in the interaction of its own parts among themselves and vis-à-vis their external environment.

The example of the thermostat was a fairly simple-minded one. Some machines exhibit much more complicated adjustments, as, for example, when a machine does not merely preserve a constant equilibrium, but performs one task after another, each one being conditional upon the completion of the preceding one. Such machines, using a feedback principle in which the machine is responsive not merely to stimuli coming from an external source, but also to the subsequent alterations which take place in itself, are analogous to an organism which has memory and learns. Such machines may even adapt their techniques of operation to the data which they acquire, so that they do not merely "learn" new information, but also whole new operational procedures, just as organisms acquire behavioral patterns and habits in the process of maturation.

One of the factors, then, which has affected the changing face of mechanism is technological advance. As machines of different kinds were developed, so the model which provided the basis of mechanical imagery for the explanation of living phenomena changed, and so the focus of attention within the domain of living things also shifted. If twentieth-century mechanists concentrate on communication systems and on organisms as transmitters of information, this is not because they deny the importance of spatial motion or of energy transfer, but only because their imagination has been captured by this particular feature which organisms and certain machines have in common.

But other factors have influenced the course of mechanism as well. One is the changing concept of the nature of matter. We have seen that mechanism is not identical with and cannot be reduced to materialism. Yet it is closely related to the concept of matter in its conviction that whatever laws are sufficient to explain the physical universe will also explain the domain of biological phenomena.

The history of science and of philosophy has been dominated by a concept of matter which may be identified as Aristotelian, because it was Aristotle (384–322 B.C.) who gave us the vocabulary in terms of which it was described. According to Aristotle, any physical object is a complex entity consisting of a passive, inert substratum which *has* properties or which is shaped into a specific form, and a formal or active element which determines the properties or, so to speak, does the shaping. Thus we may say of a table that *it* is rectangular, 36 inches high, has dimensions of 3 x 5 feet, is wooden, and has a high polish. These are all properties of the table, but what is the table? What is it that has these properties? Clearly there must be something which underlies the properties. And this something cannot itself have properties which would interfere with the manifestation of the table properties. We may then distinguish that which has the properties as the matter, from the having of the properties—the latter being what confers the "tableness" upon the matter. Matter is not tables and chairs and ice-cream cones. In itself it is sheer potentiality. It can be shaped into anything. But in order to be so pliable, it can have no character of its own apart from its receptivity to form.

Now Aristotle did not believe that there was in fact such a thing as pure matter in existence. Try to imagine what something without any form whatsoever would be like. Would it be a greasy, oozy slime? But then it would have qualities—color, density, viscosity, texture, and some shape, however variable it might be from moment to moment. We cannot conceive of something existing which does not have some properties, even though they may be difficult to describe. We speak of things being shapeless, but we do not really mean that they have no shape at all; only that the shape they have has no name, or, perhaps, that the object fails to have the shape it ideally ought to have, as for example when a dress or a person becomes shapeless. Nonetheless, we can make a logical distinction between the formless matter and the form which is imposed upon it, and this is useful as a conceptual device. It enables us to make some things intelligble, but it confuses others. It permits us to explain, for example, how a given thing can change its properties and yet remain the same.

Supposing, for example, that we paint the aforementioned table,

saw off its legs, and scallop its edges; is it the same table? Assume
that we have altered all its properties; still one can say that the
underlying matter is the same. This becomes even more significant
when we apply the same principle to organisms, for here we have
the well-known phenomenon of continuous influx and outflow of
substance. What does remain constant when you or I grow ten
years older? Although my friends may not recognize me, I will
still be the same me. It is an oversimplification of Aristotle to attribute to him the claim that it is just the material substratum
which preserves a thing's identity, but that is in part what he is
saying. (On the other hand, the doctrine of the Transubstantiation,
which is based upon an Aristotelian conception of matter, makes
just the opposite claim: that while the external properties of bread
and wine remain unchanged, their underlying matter is in fact
transformed into the body and blood of Christ. This, of course, is
a miracle, and miracles do not happen very often, but it only
makes sense if one thinks in terms of an Aristotelian conception
of matter.)

Aristotle's conception of matter enables us to understand how
there may be many instances of a single thing, for a single pattern
may be embodied in a multitude of substrates, as a single seal may
imprint a number of blobs of wax. Each is numerically distinct
from the others, but the impressed character is the same.

But note that, for Aristotle, if there is any formal organization
at all, and we have seen that there must be, since formless matter
cannot exist, then there must be an organizational principle which
is correlative to matter. Matter cannot exist unformed, although in
itself it is formless. Thus if there is matter, there is necessarily and
conjoined to it an active principle of organization. Aristotle goes
on to differentiate between things which "have a nature" and
things which do not, in terms of whether their organizational principle is imposed upon them externally, as when a carpenter shapes
a piece of wood into a chair, or whether it is determined from
within themselves, as when the qualities of gold determine that it
is gold, but not that it be fashioned into a ring or a crown. Living

things are included among those which "have a nature," but are further differentiated in that their nature is not fixed once and for all from their initiation, but is, so to speak, programmed into them so that they go through a process of development (maturation) of which the ultimate stage is the realization of their nature. Aristotle identifies this nature of the living being as its soul, and it is related to the creature as the imposed nature is to the manufactured article. Your soul is "built-in," as a house is built in accordance with an architect's blueprint.

In a metaphorical sense, then, one may attribute a soul to any physical object. The soul does not endow the object with existence as such, but it does determine its existence as a thing of the kind that it is. And, since it must be a thing of some kind, it must have a soul. My objective here is not to elaborate upon Aristotle's doctrine of the soul, but only to point out that his concept of matter as sheer passive, inert potentiality demands the existence of the soul as logically but not existentially distinct from it. This concept of matter has never really been abandoned, although it has been periodically challenged. But it has pervaded common thought, and often appears to stand immune to criticism.

In the seventeenth century, many of the Aristotelian features of matter were retained. Descartes thought of matter in terms of spatial extension and opposed it to thought as essentially passive. Even with the acceptance of the Newtonian laws of motion, the particles of matter were themselves thought to be inert but reactive to the operation of forces upon them. A variety of types of forces have been postulated—all on the premise that something is needed to bring about changes in matter or movement of matter. But invariably motion or organizational regularity seemed to require the active involvement of some factor in addition to the matter which was acted upon.

The twentieth-century concept of matter builds motion and organization right into its fundamental structure. Atoms are not hard little balls buffeted about by an external force, but are themselves complex systems including a nucleus, containing protons and neutrons, and surrounded by electrons. The negatively charged electrons move very rapidly and at varying distances from the positively charged nucleus. The position and velocity of the electrons is constantly changing, but they maintain a pattern with respect to the nucleus. According to this view, all the potential for change and organization which earlier scientists had to attribute to a principle external to matter can be understood as functions of the very nature of matter. In fact, matter cannot be described in static terms. In order to define an electron, one must refer to its distance from the nucleus and its characteristic path around it, its relationship to the other electrons, and even its individual spin on its own axis. In other words, matter as defined in modern terms necessarily includes organization, regular affinities, and change.

Modern mechanism, then, unlike its seventeenth-century ancestor, need not supplement materialism with an intelligent designer or any other purposive agent. Matter alone turns out to be sufficiently active and complicated to account for a great portion, perhaps all, of the events that take place in the physical universe.

Many mechanists now believe that matter itself has gone through a process of evolution. That brings us to still another factor which has contributed to the changing character of mechanism: the concept of evolution.

The notion that things go through a series of developmental stages is very old. There are even ancient theories of organic evolution and natural selection, which some people have regarded as precursors of Darwinism. But on the whole, in this area as in many others, the thought of Aristotle dominated the history of science, and Aristotle believed in the doctrine of fixed species. That is to say, he believed that the universe contains many different kinds of natural objects which are ultimately distinct from and irreducible to one another. Insofar as they are subject to change, this is essentially along a predetermined route of maturation. The primary task of the philosopher-scientist is to *discover* what these fundamental natures are and to state their definition, and the rules which govern their development. This doctrine is also basic to the Judeo-Christian tradition. The God of Genesis created multitudes of things which then perpetuated their own species. They are not transformable into one another. This is also the view of common sense. We see objects and organisms as distinct from one another, and we normally do not think of the cow as identical with its fleas and the bird as inseparable from its nest, even when it is feathered with the same feathers.

This is not to deny that people have been aware of the development which may take place within a species and actually may modify it. Breeders of race horses and domestic plants and animals have long known about selective breeding; but the alterations they induced were relatively minor, and surely were not based upon the conviction that all forms of life sprang originally from a common ancestor.

The doctrine of fixed species was challenged by empiricist philosophers of the seventeenth century, notably by such men as John Locke, who in his *Essay Concerning Human Understanding* gives long and gruesome descriptions of monsters which are produced when two parents of distinct species mate to produce interspecific offspring—organisms which belong to no fixed species. But the real break in the theory came in the nineteenth century under the influence of geologists and paleontologists who produced evidence of unknown and extinct species of organisms which defied classification in accordance with the classic pattern. Darwin was one of the most illustrious and persuasive of those men, and it is his doctrine of organic evolution by natural selection which eventually prevailed. Darwin's theory affirms that species originate when random variations in organisms are perpetuated by successive generations. Circumstantial conditions, such as geography and famine, lead to the isolation of one group of organisms from another, and their subsequent inbreeding then intensifies the differences between them. Alterations which fit an organism to survive in one environment might be detrimental in another, and consequently, rather different kinds of organisms tend to be perpetuated in different regions of the world. (Note, however, that there may be many alternative ways of coping with a given set of conditions, and so a variety of species can flourish in a single environment.)

Darwin himself explicitly limited the scope of his theory to organisms, and by survival he referred to the life or death of individual creatures. But the theory has been somewhat modified since his time. In the first place, it is applied to populations, rather than to individuals, and survival is interpreted not in terms of individual mortality, but statistically in terms of birth rates, longevity, and reproductive capacity. Furthermore, the concept of evolution has been expanded to cover not merely the development of organic species, but also superorganic entities such as human civilizations, as well as suborganic entities, including matter itself. This extension has been facilitated by some discoveries, such as that of the viruses, which are thought by some people to bridge the gap between the living and the nonliving, inasmuch as they have properties of both. But it is not really the uncovering of scientific fact which warrants the universalization of the principle of evolution. It is rather that this particular mode of thinking has become intellectually acceptable and that, for a variety of reasons, we no longer balk at the notion of a kind of fluid relationship between entities such that their natures can be represented as blending into one another.

Modern mechanism has availed itself of this conceptual freedom to represent life as a stage in the evolution of matter. This renders it very different from seventeenth-century mechanism most significantly in the respect that it introduces a temporal dimension which was absent from the earlier mechanism. According to the earlier view, life was conceived as a function of the structure of matter, as the consequence of a particular organization. If you shake up the pieces of glass in a kaleidoscope, you get all sorts of configurations, and if you hit the right one—BINGO— you get life! But that approach suggests a number of problems which the vitalists have rightly pointed out. When did it happen? What made it happen? Does it happen all the time? If not, why not? And if so, then why doesn't it appear to do so to us? Why can't we find and reproduce the "right" organization of matter, if that is all that is needed for life. The earlier form of mechanism does seem to cry out for some kind of mechanical engineer, and it is not an accident that most vitalistic theories have been concerned with finding that engineer in some form or other.

But modern mechanism, because of its expanded notion of machines, because of its liberated concept of matter, and because of its evolutionary approach, can by-pass some of the problems of the earlier mechanism and meet some of the challenges of vitalism. We shall now turn to a closer examination of contemporary mechanism.

CHAPTER FOUR

Foundations of Modern Mechanism

IN THE preceding chapter we considered some of the factors which have led to modifications in the historical position of mechanism. Let us next explore some of the features which are characteristic of modern mechanism. First it is necessary to make some formal observations about what mechanism is not.

Because of its historical associations with certain philosophical positions and with particular scientific doctrines, the logical role of mechanism has been frequently misunderstood. One very common misidentification of mechanism is the confusion of it with the metaphysical doctrine of *materialism*. Materialism in its strictest sense is the view that whatever is real is a form of matter or some manifestation of matter. Whatever takes place in the universe is due to some affectation of matter. In its most prevalent form materialism affirms some kind of *atomism,* according to which the universe consists of minute particles of matter in motion, and all events and processes may be understood as consequences of the interactions between these particles. In its crude Democritean form, the atoms were conceived of as solid shapes of varying density, and their interaction was limited to concussions and conjunctions. Today's atomism, as we have seen, represents matter as a despatialized locus of energy relationships. Nonetheless, all forms of materialism agree on the fundamental premise that whatever matter may be, it is the one and ultimate constituent of reality. In other words, according to materialists, it is not necessary to introduce mind or spirit or anything else in order to account for what is to be found in the universe, because everything can be accounted for as material. There is a modified form of materialism, called epiphenomenalism, according to which such outgrowths of matter as thought or spirit are in fact qualitatively different from matter. They are, however, still held to be causally dependent upon matter and not capable of producing causal results in matter. Nineteenth-century materialists enjoyed making the scandalous affirmation that the brain secretes thought, just as the liver secretes bile, or the kidneys secrete urine. According to this view, your physical state may affect the kinds of thoughts and feelings you have, but the reverse is not possible. There can be no psychosomatic disease, if by that is meant a physiological state produced by a psychic condition. Whatever takes place is entirely a physical event.

As we have indicated above, not all mechanists need be or have been materialists, but it would be difficult to be a consistent materialist without being a mechanist. The two positions are logically independent of one another; but let us try to clarify exactly what the difference between them is.

I referred to materialism as a metaphysical position, and by that I mean that it makes a claim about reality which is somehow prior to or more fundamental than our scientific or common-sense observations. How our metaphysical positions are settled upon is a very puzzling problem which we cannot take up here; but however a metaphysical stance is taken, once it is adopted, it will shape, rather than be shaped by, our scientific and common-sense observations. This is to say that, on the whole, our metaphysical commitment has priority over our scientific and common-sense beliefs such that, if challenged, they will yield to it rather than the reverse.

If one is a materialist, a number of possible modes of explanation of phenomena are automatically excluded. For example, divine intervention cannot be regarded as a possible causal force in the physical world. Certain substantive beliefs such as the immortality of the soul are also precluded. This is, of course, assuming that the soul is not regarded as simply a material entity or even a highly refined one. One does not normally arrive at one's materialism as a result of carrying out investigations with negative conclusions into such things as spiritualism and immortality; one begins with materialism, and assumes the denial of spiritualism unless required by challengers to defend one's position. In other words, materialism is held as a positive position, not as a result of failure. Hence it is frequently associated with an attitude of optimism.

Lucretius, a first-century Latin poet and disciple of the Greek materialist philosopher Epicurus, gives many arguments against the immortality of the soul. His arguments are framed within the confines of materialistic presuppositions, not in order to prove materialism, but to repudiate its opposite. Spiritualism is to be rejected on moral as well as "scientific" grounds. Following Epicurus, Lucretius represents the soul as a rarefied substance which interpenetrates and is contained within the coarser substance of the body. Then, drawing out the imagery, he says that just as the liquid contents of a jar are dispersed when the container is broken, so do the soul atoms lose their compositional integrity when the body disintegrates. The analogy makes sense only if one grants the initial materialistic premise, but this is just what a dualist would not do. Descartes, for example, starting from the premise that mind and body are two radically different kinds of substance, makes a case for the immortality of the soul precisely on the ground that if the body is destructible, then the mind, as a substance distinct in every respect from the body, must be indestructible.

I do not mean to defend either materialism or any of its alternatives, but only to indicate that such defenses are not made in the same manner as our justifications of other beliefs. There may be indirect falsifications of some metaphysical beliefs, but our experience of the world would not be substantially altered whether a particular metaphysical doctrine were true or not. It might, of course, make a difference to you personally whether you believe that the universe is material or that it is an ideal production of your own mind, but your perceptions and your knowledge of things in the world would be essentially unaffected. This is why metaphysical beliefs persist with such tenacity. They are basically irrefutable. If you believe that everything in the universe happens in accordance with a divine plan, and I believe that all events are the fortuitous consequences of blind-chance rearrangements of matter, there is very little we can do to convince one another, however coherent and plausible a case each of us is able to make for our position. But no experience can settle the issue one way or the other.

Mechanism, like materialism, is a doctrine which cannot be justified by an appeal to observation nor to any other established mode of scientific validation. But it is not a metaphysical doctrine. It does not make any direct claims about the ultimate nature of things. Like a metaphysical theory, a belief in mechanism determines how we shall interpret experience; it is not confirmed or unconfirmed by experience. Whereas metaphysics refers to reality, or to what there is, mechanism is a logical or methodological commitment. In effect, it is a statement of intention to approach the data of experience from a certain point of view. Insofar as it has substantive content, mechanism might be characterized as affirming that nature is continuous, but this is best understood as a way of saying that a single set of explanatory principles, whatever they might be, is sufficient to account for all phenomena in the physical universe. Compared to vitalism, mechanism may be viewed as a simplifying hypothesis, for it advocates the application of a single explanatory framework for all phenomena. Just as Newtonian mechanics repudiates the Aristotelian separation of terrestrial and celestial mechanics, so mechanism rejects the need for distinct explanations of the living and the nonliving. Aristotle believed that the celestial bodies were composed of a substance radically different in kind from that of the earthly elements. Hence their movement was different, and a distinct science was required to explain it. The great advance of seventeenth-century science was to incorporate these two sciences of motion into a single physical theory. The Newtonian laws of motion apply equally to heaven and Earth, to all matter everywhere.

Vitalism, like Aristotelian physics, takes the position that a single set of explanatory principles is insufficient to account for all the phenomena in the universe; specifically, an account of living phenomena requires supplementary laws and principles in addition to and not reducible to the principles and laws which account for the rest of the (nonliving) universe. Mechanism like its Newtonian model affirms that the explanation of life requires no supplementary laws and principles beyond those which are sufficient to explain the nonliving universe. We may reserve judgment for the present whether or not a mechanist may admit additional laws and principles to account for other aspects of the universe, as, for example, consciousness or thought. I should argue that it is consistent with mechanism, if unusual, to require the introduction of such "higher-level" laws. Descartes, for example, was a mechanist insofar as his explanation of life was concerned, but he regarded thought as a totally different domain of inquiry, which could not be explained in terms of the same principles.

The analogy between the mechanism-vitalism relationship and that of Aristotelian and Newtonian physics is not perfect, for Aristotle's two physics, the celestial and terrestrial, were entirely independent of one another (there being distinct kinds of motion), while the vitalist's analysis of life is not wholly independent of, but only richer than his account of the inanimate domain. The laws of physics and chemistry are not to be rejected according to vitalism, but only to be supplemented by an account of life. Nonetheless, the nature of the dispute can be illuminated by means of the analogy. The analogy also facilitates our use of another vocabulary to express the same situation—the vocabulary of *monism* and *dualism,* or *pluralism.*

The term monism will be familiar as stemming from the root

meaning *one,* as in monogamy, monocle, and monotheism, while dualism refers to *two,* and pluralism to *many.* There are a number of kinds of monism and its alternatives. Within metaphysics we may distinguish between a *numerical* monism of substance and a qualitative monism of *kind* of substance. The first affirms that there is a single substance, one entity in the universe. Not many people have held such a view, but there have been some.

The early Greek thinker, Parmenides, believed that there was one single perfect Being. Spinoza, a great Portuguese philosopher who lived in Holland in the seventeenth century, believed that there was a single Substance (which, however, possessed a number of attributes and modes). Hegel, the nineteenth-century German philosopher, believed the universe to be a single self-realizing Idea. This form of monism says there is one thing in the universe but it does not necessarily regard the nature of that thing as fixed once and for all time. (This, however, raises logical problems.) Opponents of numerical monism would say that there is more than one thing in the universe, but they might agree as to the nature of what there is. One could disagree with respect to numerical monism and still affirm qualitative monism. This second form of metaphysical monism maintains that there may be many different entities in the universe, but they are all of the same kind. Atomists, for example, say that there are innumerable atoms, but they are all of the same material nature. Leibniz, another seventeenth-century philosopher from Germany, also believed that there were many "monads" in the universe, but they were all of the same psychic quality, nonmaterial and possessing to a greater or lesser degree the power of perception. An opponent of this view might be a monist of the first kind, or he might be a pluralist with respect to the number of kinds of things in the universe. He might, for example, believe in a dualism of the mental and the material; or he might be a radical pluralist, like the fifth-century Greek philosopher Anaxagoras, who believed that there were as many fundamental kinds of "seeds" of things as there are apparent kinds of things.

Mechanism also adheres to a kind of monism, but it is a methodological, rather than a metaphysical or substantive, monism. It
makes no claims about the number and kinds of things in the universe, but it contends that they can all (with the reservation regarding "higher-level" laws noted above, p. 96) be explained in
accordance with a single set of principles and laws. It should now
be apparent how the confusion between mechanism and materialism might have arisen, for they are both monisms, and they are
not unrelated, even though they are logically distinct from one another. It may also be seen that anyone who is a materialist, *i.e.,*
who says there is a single kind of thing in the universe—namely,
matter—is likely also to be committed to a single kind of explanatory principle. Why should he need more than one? Consequently,
a materialist is most probably going to be a mechanist, but there is
no reason for the converse to be true. We have seen, for example,
that Leibniz was a mechanist but not a materialist. For similar
reasons, it is almost inconceivable, but not impossible that a vitalist be a materialist, but there surely is no more necessary connection between vitalism and spiritualism than there is between mechanism and materialism.

While, on the one hand, mechanism cannot be equated with a
basic metaphysical position, although it has certain features in
common with one, so, on the other hand, it cannot be equated
with any particular theoretical judgments about nature. Mechanism is not on a par with such laws of nature as the Newtonian
laws of motion or the Boyle-Charles gas laws.

Natural laws are, for the most part, descriptions of uniformities
of nature. They are grounded in experience and are commonly
held to be confirmed by experience. At least they are tested by experimental inquiry. It has been argued that mechanism is also ver-
ifiable, with some degree of probability, by means of empirical inquiry. If we take mechanism as affirming that a single set of
explanatory principles is sufficient to account for all phenomena
of the physical universe, and we contend further that the set of explanatory principles which is sufficient is a complete theory of

physics, then mechanism may be summarized as the view that the laws and principles of biology may be reduced to, or translated completely in terms of, a complete theory of physics (including chemistry). According to this view, mechanism would be equivalent to the denial that biology is an independent science, and the affirmation of the theoretical reducibility of biology to physics. Mechanism thus proclaims the total comprehensiveness of physics.

It is worth noting that a mechanist would not necessarily have to express himself in terms of the claim that biology is reducible to physics. In principle he need only claim that a single set of explanatory principles will account for all phenomena of the physical world. If he should prefer to translate physical principles in terms of biological ones, or to express both in terms of still another set of explanatory laws, this would not violate the conditions of mechanism. He might invent whatever name he liked for this new, all-inclusive science. Not many people have tried this approach, but some people have made suggestions which warrant bearing it in mind. Thus, the recent philosopher Alfred North Whitehead has said that biology is the study of large organisms, while physics is the study of small ones. In a similar vein, such contemporary biologists as George Gaylord Simpson have declared that the route to a unitary science is not through the reduction of all sciences to their lowest common denominator, physics, but rather through an elaboration of the principles of the most complex and inclusive science, from which the laws of the subordinate sciences could then be deduced. From this point of view one might argue that political science or perhaps cultural anthropology is the science to which all others should be "reduced."

Most self-avowed mechanists, however, take the position that a whole is explained in terms of its parts, and that if a given kind of whole, *e.g.,* biology, is decomposable into parts which are properly dealt with by another science, *e.g.,* physics, then the first science may be held to be reducible to the second. Where no decomposition into parts is possible, no reduction of the science of the whole to the science of the parts is possible. And, they say, only such a reduction would provide the method of achieving a unitary science.

Furthermore, some people believe that evidence can be given that such reduction is possible, and hence that mechanism is true. The best evidence is the fact that a number of reductions have actually been made. In recent years increasing numbers of biological substances have been analyzed into their chemical components, and some of them have actually been resynthesized out of their molecular constituents. Biochemists have studied the structure of protein, the so-called building block of life, and have even synthesized some elementary proteinoid entities. They have decoded the structure of DNA (desoxyribonucleic acid)—the storehouse and transmitter of genetic information—and have reconstituted it from its constituent elements. In a fascinating experiment with tobacco mosaic virus, scientists have even succeeded in crystallizing this lowest living form and then succeeded in reassembling it from its parts. Viable virus cells have even been produced by artificially synthesized DNA. With such successful biological reductions available, as well as similar reductions in other areas, some optimism concerning the possibility of the general reduction of biology to physics is held to be warranted.

It is also claimed that there is indirect evidence available for reducibility. The now widely prevalent evolutionary theory of matter has been offered in support of the reductionist thesis. It is maintained that if it can be shown that there actually was a time when only those entities which constitute parts of biological wholes existed, and if a mechanism can be proposed indicating how such parts could actually give rise to the biological wholes, which subsequently came into existence, then additional support, inconclusive, but persuasive, would have been provided for reductionism. Cosmologists, astronomers, geologists, and biochemists now do agree on the general hypothesis that there was a temporal succession of events during which elementary particles formed molecules, which eventually gave rise to more complex organic molecules and eventually to primitive organisms. The actual process of evolution so indicated will be discussed in greater detail later. The point to be stressed here is the methodological one that *if* such an evolutionary hypothesis can be empirically substantiated, it allegedly constitutes evidence in favor of reductionism, hence favoring mechanism.

It is, of course, agreed by all that it is impossible to certify the occurrence of a historical event which no one could possibly have witnessed. However, some people maintain that indirect support for the contention that the events did take place in the manner described can be given if the circumstances under which the events might have taken place can be replicated, and if the events can then be observed to happen. Evidence of this kind is now available in the form of a number of laboratory experiments which simulate the conditions speculatively attributed to the primordial universe. Scientists can actually provoke the formation of phenomena of the type which allegedly evolved by simulated means which might have been available in such an environment. In effect, such experiments constitute a small-scale rerun of the process of evolution, a kind of microcosmic history. Similarly, if individual entities of a more complex nature can be seen to develop out of more primitive entities, as synthesized in the laboratory, this would provide indirect evidence that the same kind of sequence takes place in the natural world. So far, little replication of this type has been achieved, but there is reason to believe it can be done.

The issue to be considered here is not so much whether the evidence is good or bad, whether it is weak or inconclusive, or whether the experiments and observations are well or poorly contrived, but rather whether such empirical inquiry can in fact constitute evidence at all for the truth of mechanism, or whether it is even relevant.

I shall argue that it is not relevant; that empirical evidence neither confirms nor refutes the truth of the mechanistic principle for the reason that mechanism is not on a logical par with theories about nature, or natural laws. Unlike them, mechanism is not even tested, much less justified, by an appeal to experience. In this respect mechanism is more like a metaphysical theory, which, as I have suggested above, is maintained prior to rather than as a consequence of experience. The difference, as we have seen, is that mechanism constitutes a methodological rather than a substantive commitment, but just as we approach experience with a prior conviction about what we shall consider as real, so also do we come with our methodological preconvictions. Whether or not we shall be satisfied with a single explanatory principle is not a discovery we make, (unless perhaps a psychological one about ourselves) but a *decision,* a plan of operation which we adopt before we actually engage in any inquiry. Mechanism is such a plan of operation, and so is vitalism. I do not deny that the adoption of one of these plans or the other can be defended, but I maintain that this is done on a variety of bases other than empirical evidence. For the present, let us defer consideration of the possible causes and reasons which might incline someone to opt for either mechanism or vitalism. Having differentiated mechanism from the metaphysical doctrine of materialism on the one hand and from particular theories of nature on the other, let us now summarize what the contentions of modern mechanism actually are.

I have characterized mechanism as a precommitment which we bring with us to any experimental inquiry. It involves the expectation that nature is continuous, or rather that so long as we are unable to account for all natural phenomena in terms of a single system of explanation, we have not yet devised an adequate system of explanation. This allows for considerable variation in the actual explanations that we adopt; and, as long as our physics and chemistry remain incomplete, we have no grounds to despair of reducing biology to these sciences. Mechanists will accuse vitalists of losing heart too easily. No doubt, the phenomena of life are extremely complex and cannot be accounted for in terms of primitive physics, but that does not preclude the applicability to them of a sophisticated and carefully elaborated physics.

J. D. Watson expresses such optimism:

We see not only that the laws of chemistry are sufficient for understanding protein structure, but also that they are consistent with all known hereditary phenomena. Complete certainty now exists among essentially all biochemists that the other characteristics of living organisms . . . will all be completely understood in terms of the coordinative interactions of small and large molecules. Much is already known about the less complex features, enough to give us confidence that further research of the intensity recently given to genetics will eventually provide man with the ability to describe with completeness the essential features that constitute life. (*The Molecular Biology of the Gene.*)

As we have seen, the concept of matter has been greatly modified by contemporary physics. Matter is no longer regarded as stuff, extended in space and subject to rigorous laws of motion. It is *despatialized, indeterministic,* and, in view of the principle of *complementarity,* its various attributes of spatial location, momentum, and velocity of motion cannot be ascribed to it simultaneously. These characterizations have been, analogously, applied to the phenomena of life. Curiously enough, it is often vitalists who first call attention to the fact that even the new physics has been forced to introduce "levels" of explanation. But rather than regard this as suggestive that as physics grows richer it will eventually encompass biology, they have argued that even as distinct levels of organization must be posited by physicists, so biology requires still more fundamental discriminations of organizational strata, and distinct sets of principles to explain their formation and character.

In part, the mechanist can accept this characterization of biology. There is no doubt that the complex combination of materials within a biological system gives rise to interrelations and metabolic interchanges with the environment which are not to be found in nonliving nature. In other words, we have among living things exhibitions of chemical and physical processes which happen not to exist in all combinations of matter. The formation of organic compounds as the result of organic processes is just one instance. But that does not imply that any adjustment of physicochemical laws is required to explain them. After all, there may well be physical processes discernible in inanimate systems which are not to be found in any living situation. But that would not appear to call for any new explanatory principles. Even within organisms, the elements show their habitual reactive properties except where other conditions determine deviations which are also describable; and the same kinetic and thermodynamic principles apply, so long as the organism is viewed as forming a system interactive with its environment rather than as closed off against its environment. Mechanistically inclined biologists proceed to test these complex interactions empirically by devising models, both conceptual and material, which will facilitate the study of these processes. Just as one can study the process of digestion by setting up an artificial analogue of enzymatic action and chemical breakdown, so one can also design communication systems which supposedly replicate our neurophysiological apparatus. Direct experimentation on such artificial models is easier than upon the less accessible system on which they are modeled.

It is the use of models which constitutes one of the bones of contention between vitalists and mechanists. The mechanists contend that a model is not misunderstood by anyone to be a literal representation of what goes on in the living system. It is, rather, a simplifying scheme which enables a scientist to concentrate upon a single process and observe how it works, thereby opening up further possible lines of study. The vitalist objects that to isolate the process, even if it is correctly identified in the first place, leads to a falsification of its mode of operation. For it does not function in isolation while contained in its organic environment, and in its proper place, it is influenced by its position within an environment. This is, of course, the view we have discussed earlier as that of the organicists, who insist upon the importance of wholes or of entire configurations. No matter what we may know of the reactive properties of an individual element, they say, this will serve little purpose when it comes to understanding how that element behaves as a part of a whole. At best, it might provide us with certain limiting conditions. The mechanistic scientist, on the other hand, admitting as he may that position within the whole is a significant factor, maintains that nonetheless, the behavior of an element can be studied as a cumulative effect of a number of factors. The point is just that there are a great many factors to be considered, and nobody claims that investigating all of them will be an easy task.

The mechanistic scientist does not regard the interactions undergone by complex organisms as *unique*. Given the appropriate conditions (which may, indeed, be rare), there is no more or less likelihood of their occurrence than there is of reactions which ordinarily occur at the inorganic level. The probability that an event will occur under specified circumstances is calculable. In some instances the probability that the circumstances will occur is itself low, although it may be a near certainty that *if* they occur, the event will take place. But these differences of probability status are in no way distinctive of organic as opposed to inorganic reactions. Any situation which is composed of a complex collection of conjoined events has a probability contingent upon the probability of each of those events occurring in the correct order. Given the satisfaction of that complex condition, the further probability of the event in question need be no more remarkable than that silver will oxidize when exposed to air, or that water will evaporate if heated to 100° C.

Convinced, as he must be, of the continuity of nature, the mechanist cannot concede that there is a probability differential between living and nonliving events except in this cumulative sense described. This is not to deny the common-sense distinction between the living and the nonliving, but only to deny its *ultimacy*. Obviously living and nonliving things do differ, but the difference may be a surface appearance, not a fundamental reality. The living and the nonliving may be describable in essentially the same language. To a large extent the distinctions that we customarily perceive are the product of self-projection rather than of observation. We extend our own self-awareness to animate nature generally, and we also interpret experience to fit our intellectual categories. Wholes and parts, as systems and levels, are not really *given* in nature. Rather, they are reflections of our thought.

This observation raises questions concerning the nature of the thought which imposes its conditions upon reality. But that takes us beyond the continuity of nature and life to the troubling question of mind—an issue which we cannot take up in the present context. Nonetheless it must be granted that it poses profound difficulties for the mechanist.

But let us assume with him that "wholeness" is not a feature of the world that needs explaining along with the chemical interaction of the elements. Then there is no need to introduce a "whole-making" factor or an organizational principle as an ingredient of the world. We may choose to make the psychological observation that we ourselves tend to condition experience by introducing "wholes" or "gestalts." The very distinctions which we labor to clarify are, then, distinctions of our own making, and biology is no more an autonomous science restricted to a unique subject matter than is a cloud somehow radically cut off from the sky against which we see it. (But note: we do see it as distinct.)

What then, one might ask, becomes of those biological laws and principles which have been accepted as usefully distinct even if not demonstrably irreducible to physics and chemistry? Many mechanists have granted that the formulation of biological laws, while not necessary on the ground of the autonomy of biology, is still profitable from the point of view of intellectual economy. Are we to deny that Mendel's rules are "specifically biological?" Surely they do not apply to anything in nature other than to organic patterns of heredity, and they do appear to apply there. A mechanist would answer this question by pointing out that such rules refer to an extremely complex occurrence, consisting of a great number of simpler processes which are systematically related to one another and to their environment. The grouping together of such a set of processes and the application to it of a conceptually unifying principle is *our* imposition upon the universe. We carve out the wholes and give them names. We identify a set of events as somehow "belonging together" and then "discover" the linking principle. This is not to say that the Mendelian rules or any other biologically useful generalizations are false. The events and processes referred to do take place; but the rule is, in effect, merely a statement of a set of initial conditions and a selected final state and an affirmation of a causal connection between them. One might view it as an abbreviated description of a huge and extremely complex sequence of events which is conceptually grouped within the confines of certain boundary conditions that serve the interest of the scientist. One or another of the same processes might in fact be included in a concurrent and overlapping conceptual framework useful for other scientific purposes.

It is, apropos of this observation, important to note that the formulation of laws specific to a particular domain of scientific inquiry is by no means unique to biology. Let us not speak of psychological laws or sociological laws, which generate the same kind of dispute and confusion as biological laws. But consider the principles of geology or of astronomy. Nobody thinks of mountains and rivers as being somehow different in nature from the matter with which physics and chemistry deals. Mountains are made up of elementary particles, and so are the most remote quasars. Yet useful principles have been formulated about how mountains are formed, and it would be not merely difficult but insane to try to translate such geological regularities into statements about atomic interactions. In principle, no doubt, every drop of water and every grain of sand and all their constituents could be accounted for, and it is, indeed, our confidence that this is the case that makes the project of carrying the reduction out appear to be unnecessary. No one seriously doubts the continuity of nature with respect to rock formation.

Yet, when it comes to biological organisms there seems to be much more room for doubt, and, consequently, when biological laws are formulated, there is a greater readiness to challenge their reducibility to physicochemical laws. And vitalists are correct in pointing out that in most cases the called-for analysis has not been provided and very likely cannot be provided. Is the burden of proof then upon the mechanist? Must he actually perform the reduction of every biological law in order to demonstrate the acceptability of his hypothesis that nature is continuous? This would surely be a Sisyphean task, and just about as useless as Sisyphus' job of rolling a giant stone up a hill only to have it roll back down again when he reached the top.

Mechanists are not such gluttons for punishment, but they have in fact made a determined effort to reduce a number of biological laws to physical principles. Much of contemporary theoretical genetics, for example, no longer operates with a hypothetical construct or explicitly biological conceptual entity—the gene—but rather expresses regularities of chemical reactions. But mechanists are just as aware of the high degree of complexity of living nature as vitalists are, and perhaps more aware of the immense number of factors internal and external to an organism which account for its variations. Mostly, mechanists are content to show that biological laws are not in conflict with physicochemical principles, and where there does appear to be a conflict, to show how such conflicts can be reconciled. Thus, for example, the argument that biological phenomena are radically distinct from purely physical ones because they violate the second law of thermodynamics can be accommodated by declaring living things to be open, rather than closed systems, deriving their energy from external sources, and therefore subject to the same principles of energy disposition as anything else in the universe.

Let me stress once more that the mechanist does not aim to prove anything experimentally, and that no amount of experimental success actually can fortify his case, for the conviction that mechanism is right is a prior condition of the formulation of the only type of experimentation that would be relevant. That conviction is that nature is continuous, that all phenomena of the natural universe are explicable within a single framework, and that where apparent discontinuities exist, they can be shown to be explicable within the framework.

One of the major areas in which discontinuity has been reputed to exist is that involving the first appearance of life on earth. To some extent we are all conditioned by popular representations of life as beginning in some vast sea or desolate landscape in the form of an egg or some peculiar organism which just appeared once upon a time. Obviously, this is a remarkable event, a break in the orderly silence of nature, and it requires some explanation. Mechanists and vitalists disagree whether the appearance of life is correctly characterized as a discontinuity in nature and to what extent one can even meaningfully speak of the "appearance of life" as if it were a single event. We shall consider the controversy concerning the origin of life in the following chapter.

CHAPTER FIVE

The Origin of Life

Spontaneous Generation

T0 SAY that life began at all is to make an apparent claim for some form of interruption of the orderly course of nature. It is to say that there was a time when no life existed and that subsequently life came into being. What were the conditions that caused this event to take place? Do they necessarily represent a break, or discontinuity? Must we, in order to explain the origin of life, if not its nature, introduce principles of explanation which are independent of and not reducible to those principles which we employ to account for the nonliving world?

ETERNITY OF LIFE

One way of evading the question of discontinuity may be to deny that life had any beginning. If life is everlasting, or even merely cotemporaneous with matter, if that did have a beginning, then the problem of explaining life's origin does not arise as a separate issue. There are proponents of the view that life is everlasting. The view is commonly defended on religious grounds, where the issue is the immortality of the living soul. But it is sometimes argued on a naturalistic basis, where the claim is not that the individual organism is immortal or that a personal identity persists after death, but only that life is by necessity living, and cannot sensibly be otherwise. Individual organisms cease to exist at death, just as they come into existence at birth; but according to the defenders of the eternity of life thesis, these facts are independent of the fact that life as such is, was, and has ever been.

They may disagree in what they take life to be. According to some views, life is an immaterial co-principle with matter. As such, *it* is everlasting, but we have *living organisms* only when in fact these two principles are conjoined—matter serving as a vessel or envelope. With such a view, one need not ask, "How did life begin?" but only how the particular combination of life and matter that we know as living individuals came into being. (It is, of course, possible to deny even this point of origin and to declare, contrary to the evidence, that there have always been organisms. But this is not an easily defended position.)

Note, however, that the above suggestion is a dualistic one. It denies the necessity of explaining the origin of life as a discontinuity of nature, but it does so by depending upon a prior discontinuity—namely, the premise that life is fundamentally different in kind from matter. Two principles of explanation are necessary, although both may deal with substances which have existed and interacted from time immemorial. Thus, the ancient Greek philosopher, Anaxagoras, believed that there were "panspermia" or germs of life scattered amongst the other "seeds" which make up the universe, and where they combined in sufficient quantity with seeds of other matter there were organisms. Similar views were held by St. Augustine, Paracelsus, Leibniz, and the French biologist Pouchet, who late in the nineteenth century argued against the position taken by Louis Pasteur on the matter of spontaneous generation. According to Pouchet, the life force never wholly abandoned organic matter, and consequently new forms of life would spontaneously generate out of decayed particles of organic substances—the presence of life permitting a reconvening of parts to form living entities.

The doctrine that life and matter are distinct, but co-temporaneous, is compatible with common experience insofar as we observe that individual organisms, so to speak, "possess" life only temporarily. According to some religious creeds, life literally passes from one body to another. A physical organism serves as a dwelling place with which the transient life element is briefly associated, passing on at the moment at which the organism is said to "die" to another physical home. But this doctrine does not account for the distinct natures of life and matter in their separate state; nor does it explain the regularity of conjunction of particular life forms with particular physical constructions. Why is it that life appears only in certain material conjugations and not others, and why are these correlations so regularly repeated? Why do animals invariably reproduce only their own kind—birds give birth to birds by laying eggs, and mammals give birth to young of their own kind viviparously? The same patterns recur without variation. If they are wholly independent, if co-eternal, one might expect life and matter to combine more irregularly.

There have also been attempts to declare life everlasting without identifying it as a substance distinct from matter, but this is a difficult position to defend. Materialists confronted with the problem of life's origin have argued that life is a constant correlative property of matter, or of matter of a certain stage of complexity. In a manner of speaking, all matter may, in this view, be said to be alive, and there is no significant qualitative difference between organic and inorganic substances. Since this is clearly contrary to our experience, the thesis has been modified: life is not an inherent *property* of matter, but a *potentiality*. Given the combination of matter in terms of certain conditions of temperature, pressure, electric potential, and so forth, those phenomena that we call living invariably arise. Strictly speaking this is not an affirmation of the everlastingness of life, but only an expression of the law of uniformity of nature. If uniform conditions prevail, then uniform consequences will follow. To specify the kindling point of wood is surely not to say that it is forever aflame, but only that given the fulfillment of those conditions, it will ignite. Similarly, this theory says only that certain conditions are necessary and sufficient to promote the occurrence of life, but that is not to say that such conditions are invariably realized.

Another argument for the eternity of life was given by materialistic scientists of the nineteenth century. They were determined to deny the theological theory of special creation of organisms, and at the same time they wished to deny spontaneous generation, which had recently been challenged by the experiments of Pasteur (1862). Special creation is scientifically undemonstrable, and spontaneous generation violates the conviction that organisms reproduce and are reproduced only by other organisms of their own kind. The solution to the impasse was to declare life eternal. But since there was evidence against its eternity on Earth and indeed against the eternity of the Earth itself, it was argued that life must have appeared on Earth from some other source sometime after the historical beginnings of this planet. Scientists were willing to consider life as having a beginning on Earth, and they construed elaborate explanations of how it might have arrived here through space; but they refused to consider the question of the origin of life as such.

Such distinguished scientists as H. von Helmholtz and Lord Kelvin believed that life, or the germs of life, were to be found eternally floating in interstellar space, and that these were carried among the cosmic bodies where, if they happened to land on a planet with favorable atmospheric conditions, they would develop and become the ancestors of the organic creatures which later inhabited the planet. As J. von Liebig described the cosmic scene: "The atmosphere of celestial bodies as well as of whirling cosmic nebulae can be regarded as the timeless sanctuary of animate forms, the eternal plantations of organic germs." (Oparin, *Origin of Life*, 36.)

Various modes of transportation were proposed. Some people

thought that primitive organisms or "cosmozoa" were carried to Earth on meteorites; others thought they were transmitted by light or radio waves. But such space travelers would have to sojourn for hundreds of thousands, possibly millions of years before reaching their earthly destination and would en route be subjected to the radically transforming effects of cosmic radiation. It is not likely that any known form of life could survive the journey. Thus, while there may be extraterrestrial forms of life—and astronomers in increasing numbers are reaching the conclusion that there are planets in other galaxies which have atmospheres hospitable to life forms like those characteristic of our own planet—nonetheless, it is extremely unlikely that there be any traffic among these forms of life or between the hospitable havens. Life may occur in a number of locations, but this is not a reason to declare its eternity. It is equally plausible that wherever equivalent material conditions exist, equivalent potentiality for the initiation of life also exists there. Hence the fact that life is to be found in a plurality of locations is not evidence that life has traveled from place to place and that it has *no* origin, but, on the contrary, that it might have originated in a number of places in identical fashion, since the prevailing conditions were identical.

Not much is gained theoretically by the declaration that life is eternal. Either it commits us to a dualism of matter and life stuff, in which case we remove one problem only by substituting another—namely, that of explaining the conjunction of these two substances; or, if we maintain that life is a natural concomitant of matter, then we cannot account for the apparently nonvital forms of matter and the long history of the Earth which predates the evidences of life.

Let us then return to the discontinuity with which we began this chapter. Assuming that life had a beginning here on Earth, how did it begin? Is the sheer fact of the presence of life on Earth a victory for the vitalist, or can this presence be accounted for within the explanatory framework of the continuity of nature?

The most flagrant denial of continuity is the doctrine of special creation—the claim that a supernatural being created living creatures more or less as we know man to create objects of his own design. Many mythological cosmologies contain such descriptions of the origin of life, among these the doctrine of Genesis. The conception of God is that of an intelligent designer who creates in accordance with a plan of his own devising. It is impossible to demonstrate the falsity of this proposal, since the deity is by definition outside the scope of empirical inquiry, and no conclusive evidence could be given against his existence or his authorship of living things. But there is also no mode of scientific investigation which would legitimize these claims. At best, we might hope to find sufficiently conclusive arguments for an alternative explanation of the origin of life, such that the doctrine of special creation may be disregarded as scientifically superfluous.

But if we discount special creation and deny the eternity of life, then how are we to account for the beginning of life? I have mentioned the doctrine of spontaneous generation as having been challenged by Louis Pasteur, but it is now evident that if life is neither eternal nor placed upon the Earth by a special act of creation, then some form of spontaneous generation must have occurred at least once. Must we then deny the findings of all those scientists culminating in the magnificent experiments of Pasteur, who labored to demonstrate that organic substrates do not give rise to living organic beings without the intervention of other organic creatures? Clearly we must not, but let us stop to reconsider what the classical doctrine of spontaneous generation involves.

The notion that living beings arise out of nonliving substances is entirely consistent with common sense and with everyday observation. Most of us have seen maggots apparently emerge from no place in decaying meat or in excrement. In unhygienic places, it is not uncommon to find worms appearing out of mud or even rats and mice out of piles of refuse. There is no reason not to regard this as a normal and natural process. Indeed, it was just such observation which led to the almost-universal belief throughout antiquity and nearly to the present that under certain circumstances living things do arise out of nonliving ones. There is, in other words, nothing "unscientific" about the belief that spontaneous generation takes place.

The doctrine of spontaneous generation is compatible with both mechanism and vitalism, and it was held by members of both groups, but its meaning was so loose that its proponents actually had very different things in mind. To declare that living organisms arise out of inert matter is only to give the barest schematic account of what does take place, and there were significant variations in alternative detailed descriptions.

We have seen that vitalists stress the discontinuity between living and nonliving phenomena, but this imposes no necessary restrictions upon the interaction of the two substances. Vitalists may regard inert matter as a necessary condition of organic life, maintaining that living organisms occur only when such matter is infused with the animating substance or force. Spontaneous generation in this view is the embodiment of the vital principle in a nonliving material substratum.

Aristotle believed that spontaneous generation occurred under the influence of solar heat as opposed to the "vital heat" required by the process of sexual generation. But heat alone does not render an organism alive. It must have a soul (or entelechy), and this, although imparted by a material medium (the sperm in the case of ordinary sexual reproduction), is itself immaterial.*

Other vitalists defended similar views of spontaneous genera-

---

* Aristotle regarded the soul as the "form" of the body, and thus in fact inseparable from it, although it could be conceptually differentiated from it just as the cube shape of a die can be abstracted from it, although it cannot be physically removed from the object. Because of the necessity of such association, it is unlikely that Aristotle intended any claims for the separate existence of body and soul, matter and vital principle. He was certainly not building a case for the immortality of the individual soul, although he may have believed in some form of collective immortality. Many vitalists are motivated to defend the immortality of the soul, but that does not appear to be Aristotle's primary objective. He did, however, hold the vitalistic belief that matter without the soul is passive and dead, and that soul, whether or not it is ever to be found in the absence of matter, is the immaterial vital element which renders a living thing alive.

tion, some even arguing that spontaneous generation constituted proof of the existence of a vital element which was under appropriate conditions conjoined with inert matter. As previously mentioned, in the eighteenth century (circa 1762–1764), an interesting debate took place between the English theologian and scientist John T. Needham and the Italian physiologist Lazzaro Spallanzani. Both sought experimental proof of their opposing convictions. Needham initiated experiments using infusions of organic materials which were heated sufficiently to destroy alien organisms and then sealed in closed containers. He maintained that, despite all precautions, microscopic organisms would appear in the infusions. This, he believed, was due to the presence of the *vegetative force* in the infusions. Spallanzani performed similar experiments, but boiled the materials at the temperature of boiling water for periods of close to an hour. He found no signs of life in the remaining infusions, and concluded that Needham's experiments had simply been inaccurate. Needham claimed to have "shown" that despite the removal of all potential procreators, the vital force had remained to produce living organisms in nonliving substances; Spallanzani made the counterclaim that with thorough sterilization and proper precautions against the survival of organisms or contamination by external organisms, there would be no spontaneous generation of organisms. Further experiments were carried out by Gay-Lussac, by Theodore Schwann, by Helmholtz, and by others, and all of these failed to diminish the prevalence of the conviction that spontaneous generation does occur. Needham responded that Spallanzani had not merely destroyed the vital force in the infused substances, but had also, through his prolonged heating of them, "spoiled" the air which remained in the vessel. In fact, the experimental procedure did remove the oxygen from the air in the container, and thus Needham's objection was not wholly unwarranted.

It is noteworthy that both Needham and Spallanzani rested their case for and against spontaneous generation on experimental evidence, and that the evidence was compatible with both their positions. This calls attention to a curious feature of science whose importance cannot be overstressed. Facts do not exist in isolated independence like marbles in a bag. They are found within contexts which are themselves determined by conventions and theories. The same observations can be understood in terms of several different interpretations; sometimes the adoption of a particular theory prohibits one's seeing at all what other people with other beliefs recognize clearly as facts. In the case of spontaneous generation, numerous experimental attempts were made to repudiate its occurrence. But the belief in it prevailed.

Prior to Pasteur's experiments in 1862, the belief in spontaneous generation was generally held by mechanists as well as by vitalists. If both the eternity of life and the doctrine of special creation are rejected, there are few alternatives to the claim that life did arise out of preexistent nonliving substances.

One alternative which arose out of a fundamentally idealistic or neo-Platonic metaphysical doctrine is the doctrine of *emanation,* or the "great chain of being." According to this view, creation is always a matter of the "greater" producing the "lesser." Since something cannot be created out of nothing, production necessarily involves the imparting of something to something else, and since one cannot give what one does not have, the initial donor must be richer in content than the recipient. That which is "more" can give rise to that which is "less," but the reverse is impossible. Assuming that living things are greater than nonliving ones (and this is an assumption which was accepted without dispute), it is then clearly impossible that life could have arisen out of nonlife, but the reverse is possible. The importance of this conviction as a metaphysical underpinning of all thought and as a direct influence upon scientific theories cannot be overestimated. It led to the retention of counterevolutionary theories long after there was substantial evidence in favor of the theory of evolution, and it led to a great deal of confusion about the nature of species. For it violates the principle of fixed species which was supported by both the Aristotelian conception of "real" categories in nature, and also the Biblical tradition of Genesis. According to the emanation theory, higher-level species could, in principle, generate lower-level ones, for whatever is present in the lower is at least virtually present in the higher. One of the glaring difficulties of this position is, of course, that it is not altogether clear, except in gross outline, what is higher and what is lower. The neo-Platonists believed that the descent went from purely spiritual to living organisms to inanimate matter. But it is not evident how living species are to be graded with respect to one another. Even if we declare vertebrates to be "higher" than invertebrates, we are left with many unranked categories within those general classes.

One concrete influence which this theory had upon scientific doctrine was as reinforcement (if not more) to the "preformation" theory held by such eminent men as Leibniz and later by Charles Bonnet. This thesis, also known as the "homunculus" theory, (see previous discussion, p. 41) maintained that the complex adult creature preexisted in the seed or embryo of an organism, and that its growth and fruition were the result of a series of "unwrappings"—not a true generation, but a development. Opposing this view was the doctrine of "epigenesis," held by such men as Claude Buffon. Their position, contrary to the emanation principle, affirmed that organisms of greater complexity could arise out of a simpler one. The controversy between the "epigenicists" and the "evolutionists" (note the special technical usage of the term as distinct from current post-Darwinian usage) was focused essentially upon embryological issues, that is, the development of the individual organism and not upon the origin of organisms as such; but it necessarily had repercussions upon other related problems as well. It was one of the major theoretical disputes of nineteenth-century science.

There is no fundamental incompatibility between the convictions of mechanism and the affirmation of spontaneous generation. As has been pointed out, spontaneous generation accords with ordinary experience, and there is no a priori reason to assume that it cannot be explained economically without recourse to the intervention of occult "vital" forces or divine principles. What needs explanation is the laws according to which the formation of living out of nonliving substances takes place, but there is no reason to expect that these laws are not discoverable and that what is now unknown is necessarily more mysterious than many other unresolved problems of nature, which produce no alarm about discontinuities within nature.

It is noteworthy that even on Biblical grounds, spontaneous generation can be accepted as a purely naturalistic phenomenon; for as the creation is described in Genesis, God bade the *Earth* to bring forth grass, the herb yielding seed and the tree yielding fruit, and He bade the *waters* to bring forth the moving creature that hath life and the fowl that may fly above the Earth. Furthermore, He created Adam out of a handful of dust. It is entirely consistent with the Biblical account of the flood and the grounding of the ark on Mount Ararat to conceive of new species of organisms coming into being afterward by means of spontaneous generation as well as by ordinary genetic means. It does, of course, remain the case that on this account God himself is required to give the command that the natural event in question take place.

The mechanist need not deny the emergence of the living from the nonliving (heterogenesis). We have seen that if he rejects both the eternity of life and special creation, few other possibilities remain open to him. But he must insist that the procedure by means of which heterogenesis takes place must be explicable in terms of ordinary natural law, or rather by an extension of those laws which are known to account for other instances of combination and production in nature. Even the claim that God is the ultimate Creator is not necessarily an obstacle for mechanism, provided that the creation in question refers alike to the animate and the inanimate. It is only when God is introduced as a special factor to account for the presence of living beings, but not for the nonliving, that the mechanist is on his guard. While the mechanist need not be irreligious, he cannot distinguish the creation of life as an exclusively religious question. Disagreements on this matter led to many of the offensives against the theory of evolution which, while not itself accounting for the origin of life, nonetheless brought that issue into the domain of science. Prior to that time, it could be safely relegated to the theologians, while scientists concerned themselves with matters pertaining to living things consequent to their origin. One eminent theologian said in criticism of Darwin, "The idea of creation belongs to religion and not to natural science; the whole superstructure of personal religion is built upon the doctrine of creation." His objection was to the scientists venturing beyond the traditional limits of their discipline.

The point here is that logically there is no reason why a mechanist should reject spontaneous generation. In fact, it is the only position open to him. But historically, belief in spontaneous generation has been vitalistic in character. Matter was regarded as a passive, inert stuff and therefore as a secondary medium to the production of life. It was a necessary condition of life, but not a sufficient one, and its activation required the presence of an additional principle, a vital force.

By 1862, it was no longer widely believed that such complicated organisms as rats and mice and frogs arose directly out of inanimate matter, but the discovery of microscopic organisms by Anton van Leeuwenhoek and others strengthened the belief that, at some primitive level of organization, spontaneous generation does occur. This conviction was once again reinforced much later when viruses, which are not even visible under a microscope, were discovered and were found to be alive. Their presence and the difficulty of accounting for it gave plausibility to the hypothesis that, in a suitably hospitable medium possibly requiring the presence of oxygen, living organisms were generated spontaneously. The unresolved question was what constituted a "suitably hospitable" medium.

In 1862, Pasteur contended that a medium was rendered "hospitable" not by vital faculties which it contained within itself, but by the minute spores and eggs of organisms which were carried about in the dust of the air which contaminated the medium. If the medium were insulated against such contamination, there would be no generation of organisms, and consequently no spontaneous generation.

Pasteur devised experiments which permitted the nutritive substance to be exposed to pure air, without dust particles, and found that here again no organisms were produced, thereby showing that it was not the air itself nor any of its components, such as oxygen, which was responsible for the production of the organisms. Conversely, the introduction of dust particles in the absence of air did promote the growth of organisms.

Pasteur also experimented with the nutrient material itself. The vitalistic proponents of spontaneous generation did not believe that it occurred indiscriminately, regardless of the host substance, but only in organic materials which retained the life force. Their view was that this force, if it had not been "spoiled," as by Spallanzani's excessive heating of the nutrient, could be stimulated to reassemble the constituent material into new organisms. But Pasteur showed that even inorganic materials, when exposed to the spore-laden dust particles, would yield organisms, while both organic and inorganic materials which were not so exposed remained barren. Later elaborations of Pasteur's experiments by John Tyndall showed that specific kinds of organisms found in the media were invariably associated with specific kinds of microorganisms in the air, and that these differed regularly in their responsiveness to different kinds of control. Tyndall found that measures, such as boiling to a certain temperature, which might be sufficient to hinder the production of one kind of organism, might not be effective against the spores of another. This observation strengthened his conviction that there was a correlation between the microorganisms in the air and the subsequent generation of organisms in the nutrient. As such, it furthered the case against spontaneous generation and also against those who argued that there were present in the air particles of continuous life stuff, "panspermia," which are the undifferentiated cause of all life.

The panspermia theory is based on the notion of unity and uniqueness of life. While species of organisms are admitted to be differentiated in a multitude of ways, it is nonetheless claimed that life is a unique property, common and necessary to them all. The presence of panspermia in the atmosphere is responsible for the generation of *living* things, not for any specifically identifying features which cause them to be a member of this or that genre of organisms. In order to account for specific differences, one must look for factors other than that which is responsible for the generation of life. But Tyndall's research made it apparent not only that bacterial strains are genetically distinct, but also that particular organisms are generated when and *only* when there is evidence of the prior presence of organisms of that same kind. In other words, life invariably takes specific forms and does so under the influence of prior occurrences of that same life form.

These findings turned out to be enormously important for practical as well as for theoretical reasons. Pasteur's own interest in the matter was evoked in the course of his investigations of ferments used by the wine industry. The differential control of bacterial growth soon was recognized as significant for the control of disease in man as well as in animals and plants. The technique of sterilization by avoiding contaminants in the air was put to use in preventive medicine by Joseph Lister. Most of today's practices of sanitation and hygiene are directly descended from these important experiments and observations about spontaneous generation.

The ingenuity and the thoroughness of these experiments left little doubt that known organisms do not arise out of lifeless matter, and with that the issue of spontaneous generation appeared to be set to rest. All living things are produced by progenitors of their own kind. Only a few dissenters remained to cast doubt, and they generally argued on doctrinaire rather than scientific grounds. Scientists turned their attention to other matters, and the case for spontaneous generation was closed.

But let us make the purely logical point that it cannot be demonstrated that something which is logically possible *never* occurs. There *are* no unicorns; but there is no reason why there *could* not be. Anything which *could* exist *might* exist, and the fact that we have not found an instance of it may just as well be evidence of our inefficiency as of its nonexistence. Except where an occurrence would be logically contradictory (*e.g.,* for an object to be simultaneously wholly round and wholly square), we cannot know with certainty that an event can never occur. This is an argument against the definitiveness of the case against spontaneous generation, but it is not a very impressive one. We do not believe that something is the case just because it is possible for it to be, any more than we believe that something is not the case just because it is not necessarily so. What the argument does show is that neither Pasteur's nor anybody else's experiments constitute conclusive evidence against the doctrine of spontaneous generation.

But have we then any reason to believe it after such exquisite evidence has been brought to bear against it? If there is nothing but a logical quibble in its favor, are we not better off disbelieving? Must we perform endless experiments with endless varieties of infusions and conditions before we are satisfied that probably no organisms arise spontaneously? Normally that would appear to be unnecessary caution. But there is a special factor to be borne in mind in this case. While it is unlikely that anyone would wish to revive the old tales that rats and worms and mushrooms, or even bacteria and viruses, arise spontaneously, we must recall the line of reasoning that led us to consider that question in the first place. We were concerned with the origin of life. If we reject the doctrine of the eternity of life and the doctrine of special creation and the claim that life is prior to and the source of matter, then what remains but some form of spontaneous generation? Must we not grant that, given the presence of life on Earth now and the absence of life on Earth at some prehistoric time, spontaneous generation must have occurred at least once? And if we grant this, then must we not grant further that, if those conditions under which life arose once were to be repeated, then, in conformity with the law of uniformity of nature, life would arise a second and a third and a fourth time? It is particularly the responsibility of the mechanist, who is committed to continuity of nature and economy of explanation, to avoid the postulation of unique events. For the identification of an event as unique is, in effect, to profess ignorance of it or to declare it inexplicable.

To explain something scientifically is to claim to understand it as an instance covered by a general law; it is to classify it as one of a kind (even if, as a matter of historical fact, there happen to be no other instances). Consequently, the mechanist, consistently with his own guidelines, is forced to conclude: (a) that life did arise spontaneously without the intervention of extraneous nonnatural forces out of nonliving matter; (b) that the procedure by which it did so is explicable in terms of natural laws describing the nature of matter and its behavior under certain circumstances; (c) that whenever those circumstances are present, the described behavior can be expected to occur; and (d) that consequently, it is at least probable that spontaneous generation has taken place not once, but many times in the universe, wherever those conditions which are conducive to life are prevalent.

Within the last thirty to forty years, scientists have come to reevaluate their own position on spontaneous generation. Under the dominating influence of the Russian biochemist A. I. Oparin, they have come to place Pasteur's conclusions in perspective as denying the spontaneous generation of complex organisms under present conditions. But they now recognize that in some earlier geologic period, under conditions different from those which are now prevalent, organisms of a primitive order must have arisen spontaneously. Currently there is a great deal of investigation of what these conditions might have been and how the original forms of life did occur.

We shall consider the contemporary theory of spontaneous generation in the following chapter, but some additional remarks may be made here concerning the historical views of spontaneous generation discussed in the present chapter. All of the views we have considered tend toward a "magical" concept of spontaneous generation. One thinks of a visibly living organism, preferably something that wiggles and squirms, emerging suddenly out of a cold, lifeless stuff. And that does seem rather remarkable and wholly improbable. But the earliest forms of life need not have been as complex as even the simplest forms of life with which we are acquainted today. All living things are composed of a surprisingly small number of fundamental substances. They contain water, certain salts of varieties which are to be found in the ocean, and compounds of carbon known as *organic* compounds. The latter are primarily composed of atoms of carbon, oxygen, nitrogen, and hydrogen in varying combinations. To produce a living thing then requires that these few elements be combined in one of the many forms of structure which possesses those features that we associate with living things. Life is biochemically economical, but morphologically extremely rich. Few primitive elements are involved, but the multiplicity and diversity of their combination is remarkable. The variations of form among living creatures is immense. But we need not even assume that all of those features which we now associate with life were present in primitive organisms. The enormous variety which is now to be found among the characteristics of living things prevents us from fixing upon any one or a few features as absolutely essential to life. Furthermore, we are aware that many of those properties now possessed by living things are the product of thousands of years of evolution. As organisms evolved, so did their environment, and consequently we may speculate that the earliest form of life which was spontaneously generated from inanimate matter was a composition of organic compounds which conceivably would not even survive in today's environment.

It must be borne in mind, in addition, that not only did this early organism have vast amounts of time in which to evolve into forms of life familiar to us now, but that it also was the product of even longer periods of time during which innumerable chemical combinations and dissolutions occurred. We must think in terms of spans of time so long that just about any possible combination of elements would be likely to occur at least once. We are dealing with a factor of probability according to which an event, whose occurrence may be highly unlikely over a limited period of time, becomes almost inevitable given a sufficiently long period of time. This is a difficult concept to grasp, but it may be illustrated by an analogy. Supposing that you had an urn which contained 1,000 marbles, of which 999 were black and one red. If you reach in without looking and draw out a marble, the chances are 1 in 1,-000 that you will find the red one. In other words, the odds are 999 to 1 against you. Supposing you put the marble back and try again, and again, and again. For each individual drawing, the chances continue to be 1 in 1,000. But as the number of trials increases, the statistical likelihood that you will at least once draw the red marble also increases until, if you keep at it long enough, it becomes almost certain that you will once draw the red marble.

In similar fashion, it may be argued that, given the age of the Earth and a limited variety of atomic elements upon it, it is almost inconceivable that all possible combinations of them should fail at some time or another to have occurred.

But there are additional factors favoring the formation of the organic compounds in question. These are related to certain observations concerning the behavior of matter. Atomic elements, because of their positive and negative electrical charges, tend to attract and repel one another in such a fashion as to form stable compounds having no charge. And, while dissolution is as natural to combinations of molecules as is their combination, nonetheless some molecular aggregates acquire stability due to structural relations which arise out of their very combination. A detailed discussion of such molecular combination can be better gained from a text in chemistry or biochemistry. The point to be made here is simply that one combination leads to another, and that spontaneous generation may thus be best conceived as a sequence of steps rather than a single coming together of elements. Spontaneous generation refers to an organic compound of some complexity which takes place over a long period of time and happens by means of a series of intervening steps. When we say that life originated by spontaneous generation, then, we do not mean that there was a single event prior to which there was no life, and subsequent to which there was life. The event itself is more correctly characterized as a series of events, each following upon the previous step—not as determined by, but as facilitated by its predecessor.

Given this interpretation, the notion of spontaneous generation retains none of its vitalistic associations. There is no reference to occult, vital forces and no assumption that the regularity of nature is violated or that any inexplicable unique event has occurred. So understood, spontaneous generation can be expected to occur whenever the conditions are right, although there may be reason to believe that the number of such occasions may be relatively low. This view of spontaneous generation is wholly compatible with mechanism and provides the basis of a nonvitalistic account of the origin of life. Let us then consider what such an account might be.

CHAPTER SIX

Biopoesis, the Current View of

Spontaneous Generation

I T HAS been said that pulling a rabbit out of a hat slowly is no less a feat of magic than pulling one out rapidly. Nonetheless, we generally feel that if we can explain an occurrence as the completion of a sequence of steps rather than as a sudden coming to be, then we have relieved it of its mystery and achieved an understanding of it. To some extent, this is what modern biologists have done with spontaneous generation. It continues to be regarded as an arising of life out of nonliving matter, but the process is a demystified, wholly naturalistic, and very gradual phenomenon.

In explaining this phenomenon, we might do well to employ a concept which is fundamental to the field of geology. This is the notion of *uniformitarianism,* which is not to be confused with the philosophical principle of the uniformity of nature to which we have made earlier allusions. The latter is a philosophical premise that the same causes will always have the same effects. It is this assurance that nature is uniform which enables us to use past observations as a guide to future expectations in all matters of fact. The principle of the uniformity of nature cannot be proven to be correct, but it is implicit in all of our thinking about the world. We cannot do without it, and we have done reasonably well with it.

[129]

The principle of uniformitarianism is somewhat less universal in its scope, slightly less fundamental to our thought, but it is no less indemonstrable. It holds that the physical processes which take place on the Earth (as well as those within the Earth) are subject to unchanging natural law, and that consequently they have taken place more or less continuously and uniformly throughout the past. It is assumed, in other words, that those processes which we now observe to be taking place, bringing about physical change in the world, were also taking place in the past, and that the present configuration of the world is the effect of the past operation of those same processes. The making of this assumption permits us to observe the present state of the world and to make extrapolations from it regarding earlier states. This would not be possible if we did not assume the uniformity of the factors responsible for those states.

The doctrine of uniformitarianism is opposed to the doctrine of catastrophism, which was widely held until well into the nineteenth century, and even now has its occasional defenders such as Velikovsky, the author of *Worlds in Collision*. Catastrophism maintains that the history of the Earth is punctuated by immense catastrophes of worldwide scope which radically altered its surface, and that there is consequently no continuity between the successive states of the world. Deluges such as that described in Genesis are typical events which would periodically transform the face of the Earth radically and would, so to speak, initiate a new phase of history. Uniformitarianism grants that there are occasional catastrophes such as floods which cover a part of the Earth, or volcanic eruptions, or earthquakes, but it denies that these are of more than local significance and that they are the primary cause of physical transformation of the Earth. The most important causes of change according to the uniformitarians are those processes which operate continuously, inconspicuously, and over enormously long spans of time. Thus while an earthquake might spew forth an occasional mountain, this is a far less likely explanation of how the majority of mountains came to be than are those minute and endless processes of movement, accumulation, and erosion which continue ceaselessly.

It will be useful to think of spontaneous generation as a uniformitarian, rather than a catastrophic, event. Furthermore, we may think of it as taking place in an environment which is also undergoing continuous and imperceptible change over immense periods of time. Geological time is almost incommensurable with our ordinary sense of human historical time; but we must think more or less in terms of a ratio of one human year to a million years of geological history. Minute alterations made continuously over that time span can produce radical effects.

While it is assumed that the laws of nature remain constant, it is not assumed that the conditions in which they are operative do. Hence, it is consistent with the uniformitarian principle that there be widely different historical phases, as, for example, before and after glacial periods, or before and after the introduction of free oxygen into the atmosphere, or before and after the appearance of life.

In order to differentiate a uniformitarian representation of spontaneous generation from the more catastrophic or "magical" version of it discussed in the preceding chapter, we will use the term biopoesis to refer to the contemporary sense of life's origin. The term biopoesis was coined by the English biochemist, N. W. Pirie, who used it to designate the entire process of development from the evolution of organic out of inorganic compounds, including the production of the first protoliving creature or "eobiont." It is to be noted that the historical controversy over spontaneous generation was not actually concerned with the origin of life as such. It was a debate over the issue of whether the forms of life *then current* could be propagated by abiogenic or nonbiological means (*e.g.,* by simple conjunction of parts). That question appears to have been settled in the negative, but the answer does not solve the more fundamental problem, namely, how did life arise in the first place? We have seen that for the mechanist, there is no alternative to spontaneous generation, but it is a notion of spontaneous generation which bears little resemblance to the traditional connotation of that term, and so we will abandon the term except to make historical references and employ the expression biopoesis to refer to the contemporary mechanistic account of the origin of life.

Before expounding upon the theory of biopoesis, let me stress once more that the truth of any account of the origin of life cannot be demonstrated with certainty. Even if we were to produce an amoeba in a test tube, we would have no guarantee that nature did not follow an alternative route, or that the route we followed would have been available under earlier conditions. We cannot do more than conjecture and make plausible cases. As Pirie has noted, we cannot expect to say, "This is how life did arise"; the best that we can hope for is to say, "This is a way in which life could have arisen," and we cannot know when, if ever, we have tried all the alternatives. In this, we are limited by our present acquaintance with living things. Even though to the best of our knowledge it is now the case that all living things include protein, we cannot be sure that this is not just the most efficient way of being alive, and that other modes of life now unfamiliar to us have been superseded by the protein form. Furthermore, it is important not to be overwhelmed by the success of one's own experiments. The stronger the case that is built for the plausibility of one account, the less inclined one is to examine the merits of other accounts—and yet they may be as plausible.

To study the origin of life, we must examine the probability of development of a particular mode of life. It is equally important to enquire as to probable primitive environments in which life could have begun. Here again we must be careful not to be misled by the success of experiments. The fact that life as we know it *could* arise in an environment is not necessarily evidence that an environment of that nature *did* in fact exist, but it tends to reinforce our thinking so. In effect, we do not know what arose or from what it arose, but we do know that some combinations can be excluded and others shown to have comparatively greater probability. In making this judgment, we are guided by the uniformitarian assumption that, while the origin of life might have covered an enormous time span and may have followed a number of distinct lines of development, it will have been subject to the same physical and chemical laws as life is today. Projecting these laws backward in time, we may infer that however different from our own the primordial environment might have been, it too will have been subject to the same laws of nature. With this assumption, we may use the findings of present-day microbiology and biology to make extrapolations concerning the environment in which the origin of life is likely to have taken place.

We have noted earlier that all known living things are built out of a relatively small assortment of chemical compounds, all interrelated and based upon a few chemical reactions. This is one of the factors which has induced scientists to believe that all forms of life have a common origin. The organic compounds, proteins, carbohydrates, fats, and so forth are all formed out of the elements carbon, hydrogen, oxygen, and nitrogen. Some of these compounds are highly complex in their structure, but the number of compounds which are possible from these elements is hardly exhausted. The structure of many compounds now found in organisms is so complex that, under present conditions, the compounds are formed only through processes which occur in living matter already in existence. While they may be synthesizable in a laboratory, they do not occur naturally by combination from inorganic beginnings. It is this fact which led people to distinguish between inorganic and organic compounds to begin with. We now know that organic compounds can be synthesized directly from inorganic materials, but it remains the case that in nature they are normally produced only by living organisms.

We are led to conclude that the complex organisms current at the present time would not have been produced in the primordial environment where there were no living things; but other compounds of greater biochemical efficiency, but less structural complexity, might have existed. Pirie has noted that biochemical sophistication tends to decrease as morphological elaboration increases:

> An organism, at the primitive stage at which it is a simple bag that contains enzymes, depends on metabolites diffusing to it and must make everything it needs from whatever materials may come to hand. This puts a premium on biochemical efficiency and adaptability. An organism such as that needs the unspecialized genius of a successful Robinson Crusoe. Along with morphologic and mechanical evolution goes a greater independence from the environment, because the organism can increasingly recognize favorable or noxious conditions and arrange to enjoy or avoid them. For a mechanically highly evolved organism, biochemical expertness loses some of its survival value. (*N. W. Pirie,* "Some Underlying Assumptions," 1958 Symposium.)

His point is that the morphologically complex compounds of today could not have been produced in a preorganic environment, but they may be the evolutionary offspring of compounds which are less complex morphologically but more efficient biochemically. What kind of ancestors might today's organisms have had, and what kind of environment must there have been to support them? In order to answer these questions, we must move far beyond the traditional confines of biology to the evidence of geology and astronomy. To a remarkable extent, the findings of these independent disciplines are in agreement. We do not know exactly how our planet was formed, whether it, along with the other planets of our solar system, was a consolidation of matter escaped from the sun, or whether it was formed as a result of the same condensation of gases and dust which formed the solar system. In any case, it would have been formed out of those same elements which form the solar atmosphere, and these are now fairly well analyzed through spectroscopic studies. The planets would have formed around solid metallic cores with gaseous envelopes gradually cooling and solidifying about them. The atmosphere of the planets would vary from one another due to gravitational differences dependent upon their comparative size and the distances between them. These variations can be calculated and corroborated in view of what we know now about the atmospheres of the planets. One could expect that because of the Earth's comparatively small mass, some of the lighter elements surrounding the Earth's core would be lost from its atmosphere and would be found on the surface of the Earth only in chemical compounds, but not in a free atomic state. Whatever free oxygen there might be would escape from the higher atmosphere to interstellar space. As the temperature of the Earth was reduced, it would solidify and would retain water vapor and other hydrogen compounds as well as oxygen compounds, but it would not contain free atomic oxygen. The formation of the Earth under such conditions might have taken place 4 to 5 billion years ago, and subsequent to that there would have been a period of "chemical evolution" of inorganic chemical compounds, eventually giving rise to the first organic compound, and ultimately to the organisms now known. The atomic oxygen which is now to be found in the atmosphere is the consequence of photosynthesis; that is, it is entirely of biogenic origin.

It is a task for biologists and biochemists to determine the conditions under which life *could* originate. The problem of determining what the *actual* conditions on Earth were at a time when life might have originated is in the domain of geology, and it is a task of entirely different dimensions. Until recently, geologists have confined their own investigations to that period for which there is a paleontological record, *i.e.,* to that period for which life itself provides the evidence, which is not much longer than a half-billion years. But under the impetus of biological advances, they have now extended their inquiries backward in time, and with the aid of new methods of dating and other technological improvements, geology now comes to the support of those hypotheses advanced by biologists concerning the origin of life. There is geological evidence that those processes which take place on the surface of the Earth, the exogenic processes (as opposed to endogenic processes which occur within the interior of the Earth and are conse-quently not affected by atmospheric changes) underwent some form of radical change, suggesting that the atmosphere of the Earth lacked oxygen up to about two billion years ago. Any living creatures which might have existed prior to that time would have to have been nonoxygen-using, but there is evidence that there were such creatures.

There is, then, independent evidence from astronomy, or astro-physics, from geology, and from biochemical speculation that the primeval atmosphere of the Earth was one in which oxygen could have been present only in compound form, as in water, but not in free atomic form. This is an extremely important conjecture be-cause the importance of oxygen to contemporary life forms, both as a necessity and as a force of destruction, is well known.

In an atmosphere without oxygen, organisms like ourselves could not survive, for oxygen is necessary to our metabolism. Fur-thermore, oxygen in the present atmosphere forms a thin layer of ozone which protects us from the noxious ultraviolet rays of the sun. No contemporary form of life could survive full exposure to solar radiation. However, even today, there are forms of life which do not depend upon oxygen for respiration—for example, the anaerobic forms, which are destroyed by the direct presence of oxygen in the air. Essential as it may be to the existence of some forms of life, oxygen is also a primary agent of decomposition. Among those organic compound formations which now take place naturally only within living systems, there are some which could occur by inorganic processes in an atmosphere which excluded ox-ygen. The very ultraviolet rays which are now destructive of life could, in fact, provide the necessary energy to produce such inor-ganic reactions. Paradoxically, the factors which are now neces-sary to life would have been detrimental to its early forms, and, conversely, the conditions under which life originated would be prohibitive of its perpetuation in its present form. We must con-sider further how, without catastrophic interventions, a set of initial conditions could have given rise to a subsequent state so dia-metrically opposed in its nature.

It is, I believe, no accident that the principal investigators of this problem were individuals whose philosophical orientation was that of dialectical materialism. Oparin, who is the acknowledged contemporary dean of investigators concerning the origin of life, was himself deeply influenced by the work of Friedrich Engels, the collaborator of Marx. Oparin quotes Engels' writings from the 1870s in which Engels advocates an evolutionary notion of the de-velopment of matter and rejects both vitalistic and mechanistic ac-counts of the origin of life as occurring by means of a sudden al-teration, natural or supernatural. Oparin himself believed that hydrocarbons, the simplest of carbon compounds, could have formed in the primordial atmosphere on Earth and could subse-quently have evolved to form protein-like compounds and then colloidal systems, which were able to undergo gradual differentia-tion of their internal organization as the result of natural selection. The significantly innovative feature of this proposal is the claim that organic compounds appeared on Earth prior to the appear-ance of life and therefore necessarily not as the product of living metabolism.

Oparin also lays great stress on the fact that life did not make its first appearance as a sudden chance combination of elements which formed a stable and self-reproductive molecule or "gene." Such an event would have been haphazard and unique, and could have occurred only under the most improbable conditions. As such, it could not be scientifically studied, and its description would be practically unproductive. Since a unique event, by its very definition, occurs only once, it cannot be verified or studied; nor would careful study of it be fruitful. Scientists are concerned with the study of types of events whose occurrences are suffi-ciently regular as to allow for calculation of their probability. His-torical and analytical study of the various stages of the develop-ment of matter can be experimentally corroborated because each step has a calculable probability and lays the foundation for the making of further predictions. Recently, scientists in increasing numbers have taken up Oparin's proposal and are engaged in studying the evolution of various of the organic-chemical substances—porphyrins, phosphorus compounds, nucleic acids, and proteins—and roles they play within living systems. The internal adaptation of these compounds to one another and to the fulfillment of the function they play within the living organism continues to be a puzzle, but it too may be fruitfully studied by application of Darwin's principle of natural selection. This requires a radical extension of the principle beyond the organic domain to which it was originally applied, but it is an extension very much in keeping with the uniformitarian commitment to continuity.

There is ample evidence that the simplest of organic compounds, hydrocarbons, could have formed and probably still do form on the Earth as well as on other planets, meteors, and even stars without the intervention of any living organisms. Once living organisms have made an appearance, the available carbon on Earth would be primarily taken up in biological processes, especially photosynthesis, and there would then be less abiogenic formation of carbon compounds. But this is wholly consistent with the evolutionary thesis that successively more effective modes of motion of matter replace one another. In conjunction with this thesis, Oparin predicted that eventually the exploitation of nuclear energy would enable man to synthesize organic substances directly from carbon dioxide and would replace photosynthesis, which has the inconvenience of being dependent upon weather conditions as well as taking up a great deal of the space on the surface of the Earth. Photosynthesis is carried out by green plants; and, as the area of the Earth which is given over to plants is increasingly encroached upon by the advance of technological society, it may well be that we shall need an artificial alternative to photosynthesis in order to assure our very survival. It remains to be seen whether Oparin's prediction will be verified.

Extrapolating backward in accordance with the uniformitarian principle, and reasoning that the primordial atmosphere of the Earth would have lacked oxygen, one may draw certain inferences about the early formation of organic compounds. It is widely held that there would have been water in somewhat lesser quantity than is now present on Earth, but sufficient to provide a medium within which organic combinations could have taken place and where they would have been protected from the direct radiation of the sun. There were also mineral substrates, particularly clays, upon which compounds could have adhered. Carbon and nitrogen would have been present, mostly in compound form, and there would also have been sulfur and phosphorus which might have had an important catalytic role. The mixture of the various elements and water is referred to by Oparin and his followers as a "thin soup." Hydrogen would have been the predominant element, and its combinations would yield molecular hydrogen, water, ammonia, and methane (H, H_2O, NH_3, CH_4). Provided with these compounds and some input of energy, more complex organic compounds could be formed.

A number of possible energy sources would have been available. The most powerful of these, ultraviolet rays from the sun, would, as we have noted above, be available in the absence of oxygen, but would be obstructed once a layer of oxygen was formed in the atmosphere. Additional energy sources would be electric discharges from lightning storms, which might have been even more plentiful in the primordial atmosphere than today, and also the energy released by the disintegration of naturally radioactive substances, *i.e.,* spontaneous nuclear reactions.

While any account of the actual conditions of the primitive Earth can be only conjectural, there have in recent years been a number of experiments which show that, within such an environment and with the hypothesized energy sources, it is possible to produce relatively complex organic compounds out of simpler inorganic ones. One of the most outstanding of these was performed by S. L. Miller in 1953. Miller devised a system of closed vessels in which a mixture of hydrogen, methane, ammonia, and water was contained. The water was boiled in a flask and circulated throughout the system, where it passed continuously by an electric spark and through the remaining compounds. The nonvolatile compounds which were formed by this process were collected in a small flask. This procedure was followed over a period of a week, at the end of which a remarkable number of organic compounds had been accumulated, including some amino acids. These are the building blocks of which proteins are composed, and proteins are common to all forms of contemporary life.

It should, however, be borne in mind that not all systems which contain protein are living. Nonetheless, further experiments with the spontaneous generation of protein have also shed considerable light on biochemical metabolism. There appears to be a correlation between the evolution of protein molecules and biological efficiency, insofar as enzymes which now play a major role in metabolism are proteins. While enzymes merely catalyze reactions which, in principle, could have taken place without them, in fact, such reactions would have taken place so slowly that in effect it is the enzymes which initiate them. Attempts to generate protein out of primary organic compounds, such as those which resulted from the Miller experiments, have succeeded in producing "proteinoid amino acid polymers" which differ only minimally from natural protein chains. These synthetic "proteinoids" resemble cells inasmuch as they form organized spherules, whose boundaries, like cell walls, are selectively permeable by osmosis.

S. W. Fox, one of the experimenters responsible for the attempts to generate proteins, is very cautious in describing the relevance of his observations to conclusions concerning the problem of the origin of life. He distinguishes between "starting" life, which he describes as "producing a cell which metabolizes and reproduces itself and its metabolic pattern" and life's beginning. Scientists may, he believes, soon be in a position to "start" life, *i.e.,* to produce viable cells; but this is no guarantee that we will know how life began. It is not implausible that our technology shortcuts some of the routes which nature actually took. But it is also possible that nature's path, unencumbered by subsequent developments, might have been more direct. At best the understanding of how to "start" life may provide us with limiting conditions upon our understanding of how life did in fact begin.

A similar contribution to our understanding of the origin of life may be derived from a study of the origin of petroleum and natural gas. Most of these products are the results of organic remains of animals and plants which have been heated in the depths of the Earth's crust, in which the large molecular compounds are broken down by chemical reactions. But there is also evidence that processes of synthesis which occur under artificial conditions in a laboratory can be replicated in nature, suggesting that complex carbon compounds, under the influence of temperature and pressure, are produced naturally out of nonorganic substances. It is then plausible that this is how they were initially formed.

Under the influence of electric discharges and then-available ultraviolet radiation, primitive carbon compounds could have formed in the waters of the primordial seas. These would then have been carried about in the water where they could have mixed to form more-complicated compounds. Sufficient experimental evidence under a wide variety of circumstances now exists to justify the belief that whatever the precise conditions of the primordial Earth might have been, they would permit the appearance of such compounds. There is even evidence that one of the chief characteristics of living things—namely, the dissymmetry of the carbon compounds—can be explained as having a nonbiological origin. Dissymmetry occurs when a molecule may take two forms which are identical in terms of the atoms and groupings of atoms which it contains, but when the atoms are differently disposed in space. Thus, a carbon atom, which has four valencies, may have a particular group of atoms attached on the right in one of its forms and on the left in another. Our two hands are frequently given as examples of entities which are dissymmetric in the manner described. Although each has the identical five fingers, they are arranged in space so that they are not superimposable, but are mirror images of each other.

In nonliving syntheses of organic compounds with dissymmetric forms, the two forms are generally found in equal proportion. Equiproportionate mixtures of dissymmetric compounds are called "racemic," and it is a characteristic observation that while the organic compounds produced naturally by living processes are non-racemic (*i.e.,* one or the other spatial form tends to predominate), those which are produced synthetically in the laboratory are racemic (*i.e.,* there is an equal distribution of the alternative spatial forms). This phenomenon is also designated by the expression "optical activity." Optically active compounds are those which contain dissymmetric substances in a nonracemic mixture. Biologically produced organic compounds are thus optically active. It has been observed that protoplasm produces and stores only one antipode of dissymmetric molecules.

It was long believed that all instances of molecular dissymmetry were of biological origin and that no nonorganic instances of dissymmetry existed. It is as if all normally generated human beings had two left hands, while all artificially produced robots possessed an equal mixture of left and right hands. This distinction between natural and artificially synthesized organic compounds has long been known and was for some time regarded as a defining feature of life and associated with a vital force thought to be responsible for the discrepancy. Pasteur sought to explain the phenomenon nonvitalistically by attributing the dissymmetry of naturally formed organic compounds to asymmetric physical forces such as the direction of the rotation of the Earth or terrestrial magnetism. His attempts to prove this hypothesis experimentally were unsuccessful, but it has since been shown that instances of nonracemic mixtures of dissymmetric compounds do occur in inorganic nature and that, furthermore, it is possible without any intervention by organisms or their products to obtain such mixtures from mixtures of equal distribution. This has been done by means of ultraviolet irradiation, which is interesting since it suggests that the distribution of dissymmetric organic compounds might have arisen on Earth under primordial conditions before there was any sign of life at all.

There is an increasing number of experiments with "life substances," which are being synthesized under simulated prebiotic conditions. Even with tremendous variations, there appear to be distinguishable patterns. One of the most important of these is the pattern of natural selection. This involves an extension of the Darwinian concept from the organic to the inorganic domain, but the move appears to be justified in that the logical pattern of natural selection is maintained. Substances which can reproduce themselves do so, and with remarkable fidelity to form, but occasional variations occur on the level of organisms, where the fidelity breaks down. As in the case of organisms, these mutations frequently turn out to be either noxious to the compound in that they weaken its stability or render it more easily destructible, or they are favorable to it, enhancing its likelihood to maintain itself against competition. Presumably some accidental variations of this type led to the transition from what has been referred to as "chemical" evolution to "biological" evolution.

It has been suggested that these modes of evolutionary process are in fact carried out by different kinds of molecules, the first being a process of autocatalysis, whereby the system of molecules promotes the production of more of itself. It is characteristic of biological evolution that two different but coordinated systems of molecules are involved, the one being essentially an information-bearing system which directs the synthesis which takes place in the other system. We have two such systems interacting in the nucleic acids which provide genetic information and the proteins which build it into a physical structure. According to the evidence found in geochemical investigations, the transition from chemical to biological evolution might have taken place some 3 billion years ago, and therefore, one might expect to find some significant molecular alterations at that period. What is sought under the rubric of evolution of life is not an increase in substances, but an increase in information transfer and storage potentiality of substances. This notion of information complexity and reproduction is so crucial to the understanding of evolution at this level that J. D. Bernal has said of life that it is "essentially . . . a matter of the growth and self-complication of the informational aspects of the potentialities of matter." The process of storing and replicating information requires expenditure of energy; hence, the evolution of life entails a corresponding evolution of energy source availability.

A chemical reaction can be induced by means of an energy input from an external source. Presumably, solar radiation or lightning sparks would have provided such inputs in the primitive "soup." But organisms as we know them today have built-in stored energy which is more than sufficient for self-maintenance. This is why organisms serve as food for one another. They are, in effect, fuel supplies. The evolution of life is thus closely related to the evolution of increasingly efficient modes of accumulating, housing, and diposing of energy.

The forms of life which are most common today engage in one of two modes of accumulating energy. Most animals respire oxygen which is necessary to their metabolism. Most plants use light energy from the sun to maintain their growth. One of the by-products of this photosynthetic process is oxygen, and it is believed that all the oxygen now available in the air to oxygen-breathing organisms is derived from those organisms which engage in photosynthesis. But both photosynthesis and respiration are too complex to have been available to primitive organisms. There are, however, even now some forms of life which require neither direct oxygen nor light and which carry on functions characteristic of both plants and animals. Such organisms derive their metabolizing energy from the process of fermentation. This involves breaking down organic molecules out of a surrounding environment and rearranging their parts. It does not generate new organic materials; on the contrary, it leaves a residue of waste inorganic materials. But it provides the organism with sufficient energy to live and grow and reproduce, so long as its environmental supply lasts.

Presumably, a primal organism, which at this stage would be no more than an ordered aggregate of organic compounds, could have survived and perpetuated itself by some ancestral process of fermentation. But eventually, if no other source of organic compounds existed, it would be reduced to breaking down its own offspring and would deplete the supply of organic materials. Conceivably, a number of such dead-ended life histories did take place, but eventually the process must have been superseded by a more efficient mode of gathering energy.

This is evident, since an essentially nonproductive sequence cannot be endlessly perpetuated. Imagine a group of people nourished exclusively by cannibalism. For a period of time they could survive by frequently reproducing and eating their offspring. But this would be an exhausting and wasteful mode of subsisting. Eventually, they would become incapable of reproducing fast enough, and, having depleted the available population, they would die of starvation. But if this group of people could learn to manufacture and retain food without destroying each other, they could hope to survive. This is the advantage of photosynthesis.

It is not entirely clear at what point in the sequence of chemical evolution one would identify a being as living. Presumably the production and concentration of organic molecules within the primal "soup" would not warrant regarding it or any portion of it as living. However, as aggregates of these molecules collected, either due to their adherence to clays and mineral deposits or as a result of mutual interactions, they would form vague patches of relatively constant composition which might cover large spatial areas. Bernal has referred to these as "sub-vital units," not yet clearly defined as organisms, but having indefinite and shifting boundaries and the ability to merge and fuse with one another. At this point, there would be no real mechanism of reproduction, but only a process of subdivision and fusion.

This stage would be followed by the formation of more-structured collections of molecules which would cohere together in the form of spherical or elongated globules. These could operate as self-subsistent systems, maintaining a certain intact integrity even in the absence of an external membrane and yet also undergoing exchanges with the environment. While they would lack any reproductive mechanism, they might still possess sufficient order to qualify as Pirie's transitional being, the "eobiont." Subsequently, a genuine reproductive system could arise with the coupling of nucleotides which could govern the structuring of protein molecules in accordance with characteristic patterns. At this point, the organism is not merely breaking down organic compounds into their constituents and utilizing the parts, but is actually directing the construction of new compounds. The regularity and fixity of such combinations would itself be a matter of evolution eventually producing the nucleic acid–protein mutual-synthesis relation which is common to the present forms of life.

Photosynthesis as we now know it involves the absorption of sunlight by the chlorophyll or pigmented substance of plants. This energy is utilized to convert water and carbon compounds into carbohydrates, which are stored as a source of additional energy for the organism itself and for others which may use it as food or fuel. In the process of this conversion, oxygen is released into the atmosphere, and this oxygen then becomes available to other organisms.

It is evident that the photosynthetic process as now familiar could not have occurred in a primitive atmosphere which lacked the appropriate compounds. But an inorganic analogue of photosynthesis using sunlight and water and some mineral compounds and the leftover products of the protofermentative process could have taken place and would be responsible for the gradual appearance of oxygen in the atmosphere. Thus, the processes themselves and the environment within which they occur are mutually adaptive, and the evolution of the one draws with it the necessary evolution of the other.

Once oxygen is present in the atmosphere, a much more efficient system of energy acquisition, storage, and hence, disposition, becomes available—namely, that of respiration. Having the resources at hand, the organism is also able to develop more efficiently in other respects. It can, for example, develop more elaborate respiratory and digestive systems and more effective modes of reproduction. At this point, there is no longer any possible doubt that we are dealing with a living organism which perpetuates itself with relative constancy through interchange with its environment, but it is also clear that the distinction between this stage and previous ones cannot be defined except arbitrarily. Fermentation yields conditions conducive to photosynthesis, whose products condition the atmosphere to render it hospitable to organisms capable of respiration. It is then easy to comprehend how the concept of natural selection, employed by Darwin as a mechanism to explain evolution on the level of organic development, could also be extended to account for preorganic and even prebiological evolution.

It should be borne in mind that the conditions of life, which the proponents of the biopoesis theory describe, are represented as applying to the Earth, where there is evidence of a particular type of atmospheric environment. Given different conditions, we cannot assume that life would not occur, but we must assume that it would occur differently. Speculation along these lines has revived the claim, long the source of ideological and religious controversies of the past, that there may be forms of life elsewhere in the universe which differ from those familiar to us in being adapted to their own environment.

But such differences would not be grounds for the repudiation of uniformitarianism. On the contrary, their alleged existence is based upon the premise that uniformitarian conditions hold and that projections and extrapolations concerning chemical combinations of the past and in faraway places can be made on the ground of our present knowledge of chemical combinations today. The doctrine of biopoesis does not treat the appearance of life as a unique, inexplicable event. In that sense, it is not spontaneous. It is maintained, rather, that, given the properties of matter, certain conjunctions are more stable than others, and these, if they occur, are more likely to be perpetuated than dissolved. Once conjoined, they lay the groundwork for new orders of stable conjunction. While there is nothing inevitable about these combinations, their comparative probability is a matter of calculation, and the presence of certain conditions will be conducive to their realization. In principle, the effect of other conditions ought also to be calculable so that some conjecture about alternative forms of life is possible. In fact, there are speculative accounts of life in an atmosphere which contains ammonium but lacks oxygen.

Much remains to be learned about the spontaneous generation of life on our own planet, let alone the possible evolution of life forms in atmospheres other than our own. But the topic which once was a stronghold for religious controversy and mystical disputation now falls squarely within the domain of scientific investigation and rational discourse. Nonetheless, the issue is not without its philosophical uncertainties, and we shall conclude our discussion of the nature and origin of life with a consideration of some of the philosophical issues which are raised by the present analysis.

CHAPTER SEVEN

Philosophical Issues

O NE of the beauties of philosophy is that philosophical questions arise everywhere and at all levels. We have already touched upon a number of philosophical problems in the course of the preceding chapters: such issues as the relation between mechanism and materialism, the problem of defining "life", and the connection between theological convictions and vitalistic commitments. These, as well as a number of the other issues which have occupied our attention, are more frequently matters of concern to philosophers than to practicing scientists. However, the distinction between scientific and philosophical questions is not always clear-cut, and is in any case somewhat arbitrary. We cannot hope to cover all the philosophical questions which might present themselves in the course of an investigation such as the one upon which we are embarked. These will vary with the point of view of the reader and the writer, for one man's philosophy is another man's common sense or perhaps his nonsense.

Nonetheless, biological inquiry almost inevitably evokes certain questions of a broadly philosophical nature. In the preceding chapter, for example, we were left with the suggestion that the occurrence of life might well be regarded as *inevitable; i.e.,* that given certain features of matter, specifiable by the physicist, and given certain environmental conditions specifiable by geochemistry and astronomy, those combinations that we designate as living and associate with life *must* occur. This being so, we might expect that our inquiry will be a simple experimental one. Merely reproduce those conditions and then observe if, in fact, living creatures are formed. We have seen, however, that the creation of life, to say nothing of the study of its origin, is not such a simple matter. Spontaneous generation does not happen every day. If it did, it would be a timeless phenomenon analogous to the crystallization of a solution and would clearly fall into the domain of physics proper. But life is a historical phenomenon. The combinations described are themselves the products of evolutionary experimentation. However accurate our replication of the primordial conditions, we cannot replicate the billion or so years of random combination which preceded the appearance of life, nor the sequence of naturally selected combinations which led to that form of life with which we are acquainted. The claim that life is inevitable must be taken in a special, qualified sense. It does not refer to a uniformity that may be observed under recurring circumstances. On the contrary, we have seen that its singular occurrence in fact precludes the recurrence of those very same circumstances. For historical processes are irreversible. Once an historical event has taken place, the circumstances within which it exists are sufficiently altered to render it impossible that the identical event occur again.

The ancient Greek philosopher Heraclitus said that one cannot step into the same river twice, for in the interim both the pedestrian and the river will have undergone change. He was corrected by his pupil, Cratylus, who declared that one cannot even step in the river once, since the flow of water during the beginning and end of the step prevents it from being "the same" river. This may be an exaggerated point of view, but it is fruitfully applied to the issue of historical processes. An historical event cannot be repeated. The second production would differ from the first, not merely in the sense of succeeding it in time, but also in that the first occurrence brings about changes, however slight, such that the surrounding conditions of the second cannot be identical with those of the first. If nothing else, the environment of the second event includes the fact of the prior occurrence of the first. We must then reconcile the notion of life's inevitability with the apparently incompatible notion of its historical uniqueness. How can life be a phenomenon whose single occurrence is both inevitable and historically unrepeatable?

Speaking of natural, rather than logical inevitabilities, we are inclined to think of regularly occurring sequences of events which, following the principle of the uniformity of nature, fall into the same sequential pattern. What is meant by referring to them as inevitable is precisely that the causal pattern does repeat itself. Events of type A are regularly succeeded by events of type B; hence any particular A event is expected to be followed by a B event. Yet with respect to the occurrence of life, we affirm the inevitability of the event and, in the same breath, deny the very possibility of its repetition. We must refer back to an earlier discussion of probabilities (Chapter 4). There we noted that the calculation of the probability of life was not a matter of assessing frequencies. We cannot know how many times it happened or how many trials preceded it. But we can affirm that, given an infinity of time and no adverse conditions to prevent the occurrence of a combination which might logically take place, it becomes virtually a necessity that the combination will take place at least once. And, having occurred once, it will not take place again. It must be noted, further, that the inevitability to which we refer is not, strictly speaking, the inevitability of life, or of any particular form of life, but only of a particular organization of matter which, in turn, is succeeded by similarly inevitable organizations—each possible, but not necessary in their own right, and each enhancing, but not entailing the likelihood of the next phase.

It is not necessary, for example, that a living organism possess one or another specific form of locomotive adaptations, but given the capacity, *e.g.,* to propel itself through a liquid medium, it is more probable that an organism would survive in water than that it be found in desert areas, where other locomotive organs would be more effective. And, given the fact that the organism is water-bound, the probability is increased that certain other aquatic adaptations (*e.g.,* relating to shape, breathing apparatus, nutritional sources, and so forth) will also be selectively acquired. One cannot predict at the outset what properties a species will possess; but the possession of certain properties limits the range of additional properties which might be expected.

The inevitability of any particular form of life is contingent upon the events which precede and surround it. Life could not have been predicted; nor can we say with any certainty what forms life would take under conditions dissimilar from those under which it did arise. Some attempts have, in fact, been made to speculate on the nature of life in an atmosphere chemically different from our own, but these can be only speculative and schematic projections of our knowledge of chemical possibilities, and cannot account for the statistical fluctuations and the vagaries of natural selection, which would determine the character of such beings. At best, one can say that in other universes, where chemical conditions are like those of our own, the inevitabilities are the same, and life if it is present would take a form very much like our own. In an atmosphere without oxygen, or one where another element, such as ammonium, played the central role which oxygen plays in our atmosphere, one might conjecture what sort of ammonium-using creatures might be found. But the notion of an ammonium-using organism requires so many alterations of our chemical expectations of both the organism and its environment, that no realistic projection of what such a creature might be like is possible. Astronomers believe that there are worlds where the chemical conditions are fundamentally like those of our own planet, and that consequently there may be thousands and hundreds of thousands of planets like our own where life as we know it may also exist. This appears to them far more likely, or at least more within the realm of speculation, than conjectures about the possibility and likely nature of life *within* our own planetary system, where the chemical atmosphere is known to differ from our own.

Our discussion of life is necessarily limited to life in its terrestrial form. While there can be no certainty of the unity even of this form, it is generally believed, on the basis of the paucity of chemical combinations out of all those which are available, that life on Earth arose as a consequence of a single series of chemical conjunctions. It is possible that there were a number of abortive forms of life, or even that there are now formations of chemical combinations which, in a primitive environment and in the absence of predators, might also have developed into variant life forms. It is, however, one of the curiosities of life that, once initiated, it becomes not merely self-propelling, but also self-destructive and self-annihilative. Chances are that not more than a single model of life would survive such competition.

The destructiveness of life is familiar to us on the everyday level at which we exist. We are acquainted with the problems of overpopulation, with natural aggressiveness, and with the predator and the prey as related by the food chain, but there is an endemic life feature much more fundamentally destructive than all this, and it is found even at the most primitive level of biogenesis. We have noted that oxygen is introduced into the atmosphere through the process of photosynthesis. In the absence of oxygen, solar radiation and electric discharges stimulate the production of chemical compounds, which are no longer formed once a layer of oxygen forms between the Earth and the source of energy. The presence of oxygen thus puts an end to this primitive form of life production. Oxygen is also a more direct impediment to life as a result of the corruption it causes in the form of oxidation of inorganic substances and organic rotting. Furthermore, the increasingly complex chemical compounds, which become possible as a result of the presence of free oxygen, are antithetical to and destroy one another. Biological efficiency is purchased at the expense of the less efficient forms, which are destroyed by the new complexes.

Remarkably, this tendency is counterbalanced by a phenomenon which arises with the increased efficiency of the compounds. Chemical reactions which occur extremely slowly in a primitive environment would be prohibited from taking place at all in an increasingly complex and competitive environment. But among the complex compounds which are formed are enzymes—protein compounds which serve as catalysts enabling those very reactions to occur more rapidly.

At each phase of development, new facilities arise, but so do new impediments to the perpetuation of the prior forms of life. Greater specialization of organisms leads to greater efficiency, but also to a decline in self-sufficiency. Among the capacities that arise as an organism becomes more specialized, more dependent, and more individuated is the capacity for death.

In its most primitive phase, matter is truly deathless. Molecular combinations form and are dissolved. But in accordance with the principle of conservation of matter, the constituent parts do not literally cease to exist. In fact, we do not speak of death at all as referring to sheer destruction or dissolution (unless we speak metaphorically), but only as a correlative of life, or, more correctly, as its dialectical counterpart. If death were simply a breaking down or dissociation of matter, we would not discriminate between the inanimate and the dead, but we do regard as dead only those substances which have actually been, or might have been, alive. While a dead organism reverts, as it were, to a state of inanimate being, ceasing to engage in those processes which are characteristic of life, it does not thereby become an inanimate object. We reserve the term "inanimate" to refer to those things which never were and never could be alive. And while a dead organism behaves like an inanimate one in all significant respects, we nonetheless honor its history by continuing to regard it as not identical with things which never were alive. To call something "dead" (literally) is to make a historical claim. It tells us that the object once was alive, and that this state is now terminated.

It is, furthermore, worth noting that while not all forms of dissolution of organization are identified as death, the notion of death is commonly associated with some form of dissolution of dynamic organization. Disorganization appears to be a necessary condition of death, and, causally, a sufficient one, but unless the organization is coupled with additional self-stabilizing and integrative features, we do not regard an object which is simply disorganized as dead. This is one reason why the most primitive forms of proto-life, sometimes characterized simply as a "bag of enzymes," are not thought of as dying. It is clear that we cannot formulate a satisfactory conception of death so long as we lack a clear notion of life, but some ideas may have been suggested by our discussion of biopoesis. In the primary phase of biopoesis, molecular combinations occur simply as a consequence of mechanical and chemical circumstances. Whether or not a compound remains intact is also a consequence of physical and chemical factors, which have nothing to do with the individual characteristics of the particles which are conjoined. We have statistical knowledge that the class of x atoms reacts in a given fashion with the class of y atoms. The individual variations of an x or y atom, if there are any, are irrelevant to this reaction. But as the combinations grow more complex, they begin to vary from one another in small details, and these become the basis of larger variations in the reactive patterns of the compounds. With the increase of complexity comes an increase of individuality. Until entities are individuated from one another, we do not refer to their destruction as death.

Death is associated with complex entities. This being the case, the death of an entity involves a change in its complex state, but not necessarily the destruction or even the removal of any of its constituents. Conversely, it is possible for a complex component of an organism—a cell, for example—to "die" without thereby incurring the death of the whole of which it is a part. The death of the whole organism involves a change of total configuration, but, again, not merely a change as such. It is not at all clear when an organism is properly taken as dead. Not all changes constitute or entail death. Evidently some minute alterations are more significant than other more visible ones. There is an obscure form of immortality which is achieved by many of the lower forms of organism through the process of self-replication. Such organisms, despite change that amounts almost to dissolution and recombination, cannot be said to die at all.

Immortality has long been one of the most perplexing philosophical questions. There is hardly a philosopher of any repute who has not at some time addressed himself to the question of whether or not some portion of man survives his physical death. Much of the speculation on this matter is motivated by a compelling need to think of one's life as having some meaning, as being more than an absurd contingency in a mindless universe. Somehow, the prolonging of personal existence seems to answer this need. Those who defend human immortality generally do so on the basis of some form of dualism, in accordance with which it is possible to maintain that the body undergoes decay and destruction, but the soul remains alive. Curiously, our present discussion suggests that it is matter which is indestructible and which is passed on from generation to generation. The immortality possessed by the simple unicellular organism is not that of an individual, but is rather that of the material continuity perpetuated in its offspring. Indeed, it appears to be the case that, as the individuality of organisms increases, their mortality also increases. That which is deathless is also undifferentiable from other beings like itself in structural organization and disposition of matter; but that which dies is distinguishable from other beings in virtue of the possession of those very features whose dissolution marks the death of the organism.

We may illustrate this point with a simple experiment using drops of water upon a surface such as that of waxed paper. Suppose we tip the paper, running the drops together into a single drop. No water molecules have been destroyed, yet the separate drops have lost their individual identity. If we now break up the conjoined drop into a number of drops equal to that of the original collection, we have the same molecules arranged in the same number of distinct entities, but we have no way of determining whether in fact we have "the same" drops. In a similar manner, some viruses conjugate in pairs, exchanging genetic material. When they separate, they continue to exist as independent organisms, but they are no longer "the same" as before their conjunction. On the other hand, they have not ceased to exist; they continue to be organisms of the same type, and their material constituents have been neither augmented nor decreased. Regarded as individuals, they no longer are what they were; yet since they continue to function as distinct individuals, it is not correct to say they have died. In this sense, the lower organisms do achieve a kind of immortality which more complex organisms lack. Perhaps it is because we do not attribute individual identity to the lower organisms, and hence think of them as having none to lose, that we are ready to regard them as immortal in these terms. Regardless of the material perpetuation of higher organisms, we would grant them immortality only on the ground that a specific and unique identity is preserved.

We refer figuratively to the immortality which people achieve through their children and grandchildren. By this, we intend the ideas and values, which have been transmitted, as well as the memory, which children bear of their parents, and we tend to think of these as representing the enduring uniqueness of the ancestor. His personal identity is preserved through the persons of his descendants. What is literally transmitted is not these individualizing features, but just the contrary: the impersonal and interchangeable material components. A one-celled organism is immortal precisely through the fact that there is nothing in particular which identifies it as this or that organism, hence it persists so long as its substance and the pattern of its organization persist. A complex being differs from all others, and its dissolution marks the end of the being of something which is not replicated even if its constituent parts are preserved.

The organism which reproduces itself by division, creating two (or more) exact replicas out of its own substance cannot be said to die, for all its properties remain intact except its numerical identity. It is thus only in an abstract sense that destruction occurs and that the parent exhibits mortality. Unicellular organisms such as amoebae or paramecia normally subdivide by mitosis, and the newly formed pairs of cells are referred to as "daughter cells." Their constituents are literally those of their parent cells, divided and replicated. The parent cell is thus, so to speak, dismembered and reembodied in its offspring. There is no corpse of the parent cell to be found, and yet, while it has not died, as an individual cell, it has ceased to exist. In higher organisms, a much more concrete destruction of the parent occurs, and in some instances, parental mortality is the price of reproduction. We are acquainted with the forms of organisms which reproduce at the cost of the life of one of the parents. In some insects and spiders, for example, the eggs are laid in the body of one parent which will serve as food for the newly hatched larvae. Other animals, some fish, turtles, and amphibians, for example, die immediately upon performing the act of copulation, and in organisms such as man, there is a gradual degeneration once the reproductive period is past. It is, then, more than a poetic notion that children are the embodiment of both our mortality and our immortality.

It is noteworthy that the demarcation of life and death becomes an increasingly serious practical problem as our ability to prolong life increases. It is the practice for legal and medical purposes to distinguish between "biological" and "clinical" death. The latter refers to an arbitrarily defined state when the organism as a whole is no longer fully functional, although some number of its component parts—tissues, organs, and cells—may still he operative. The definition of this point has preoccupied the medical profession of late because of its significance for transplant operations. Transplantation of vital organs, such as hearts and lungs, are made from the bodies of "clinically dead" but biologically alive patients into other patients who are clinically alive, but who have vital organs which are nonfunctional. These operations must be done immediately upon the clinical death of the donor, for the biological life of the organ depends upon its being integrated into a living system. This suggests that both life and death are systemically defined. Clinical death refers to a major breakdown of the entire system which may yet include viable subsystems, but one cannot count on it. The total organism may be regarded as a system of systems which can tolerate a certain amount of internal disruption, but which collapses when such damage exceeds a given limit. The designation of that limit defines clinical death.

The systemic approach to the problem of life has been adopted by a number of scientists and philosophers, who view it as falling within a broader science of self-regulating mechanisms. The study of these mechanisms is called "cybernetics" as proposed by one of the major theoreticians of systems theory, Norbert Wiener. Wiener derived this term from the Greek word *kybernetes,* meaning "helmsman," referring to the purposive, or self-governing capacity of some mechanisms which have the ability to direct their own conduct, or even to redirect it in such a manner as to maintain a steady course. Interest in such mechanisms generated a whole new discipline of "communications engineering" which studies the transference, storage, and utilization of information. Cybernetic mechanisms have the capacity not merely to assimilate and process information from the external world, but also, by means of "feedback" features, to receive and record information about their own operation, and to adjust their own rules of operation on the basis of this incoming data. This ability has been likened to the ability to remember and learn. The science of cybernetics developed as self-regulative mechanical devices became increasingly complex and, in their sophisticated adaptation of means to ends, increasingly similar to the self-regulative behavior of organisms. The development of such "goal-directed" systems as automatic spaceship controls, guided missiles, automated factories, and so forth has suggested a new model for the understanding of living organisms in mechanistic terms.

Traditionally, one of the major arguments given in favor of vitalism has been that, since living organisms alone exhibit purposive behavior, they must be accounted for in terms fundamentally different from the principles which explain the behavior of nonliving creatures. It is, of course, evident that nonliving things may be designed for the purpose of achieving specific objectives. From the beginning of man's history, he has used objects and shaped them to serve his own needs. In this sense, a stone axe or an automobile exhibits purposive behavior. Their purpose is built in, and they will, under normal conditions, function to fulfill that purpose. But objects of this type have no inherent adaptability. Under unusual conditions, if their regular mode of functioning is inhibited, they will simply become inoperative. They cannot adjust to a change in conditions unless they are adjusted, again by an external designer.

Organisms, on the other hand, are characteristically plastic. If one avenue of achievement is blocked, they can adopt another. In effect, the dynamic equilibrium commonly associated with life consists precisely in the constant adaptation over periods of millions of years of the organism to environmental conditions which impose constantly varying constraints upon it. This adaptation, said to be "in order to" achieve an aim is often referred to as purposiveness. This is not necessarily intended as an attribution of conscious intention to organisms which cannot clearly be held to be conscious at all. While we may have doubts about some of the higher mammals, there is very little temptation to claim that protozoans or plants are conscious. This is not to say that plants and lower forms of life do not react to stimuli. Plants are notoriously responsive to light, and by some accounts to sound. And some plants, known to nurserymen as "sensitive plants," visibly shrivel when touched. But we do not attribute to these creatures an inner awareness of themselves as reacting. It is such self-consciousness that is presupposed by the capacity to act purposively.

Nonetheless, these lower forms of life may be observed to act *as if* they were under the guidance of specific aims. Like the most intelligent of creatures, and like ourselves when we are conscious of particular objectives, they vary their behavior to fit the circumstances and to satisfy their needs. Purposiveness, then, does not necessarily entail consciousness. Rather, it refers to a pattern of behavior, which is not merely goal-directed, but which is also dominated by the goal in the sense that it is modified as the goal becomes more or less accessible or as impediments interfere with direct access to it. Indeed, behavior may be so modified by its purpose even when the goal is absent or the objective unfulfilled. One's goal, for example, may be to find a nonexistent ideal, the dream of which becomes a consuming passion. Or, one may set out to achieve a task whose proportions one has misjudged and which is never accomplished. Races unwon are still run for the purpose of winning, and cries unheard may still be cries for help or for attraction. A search for a lost object is no less purposive if the object remains unfound. All such behavior, whether conscious or not, whether successfully realized or not, is referred to as "teleological," deriving from the Greek word *telos* meaning "end" or "aim."

The explanation of teleological behavior has long been a problem of interest to both philosophers and scientists. Logically, it poses problems because purposes are future-oriented, implying an awareness of the end to be achieved, when that end is not yet in existence. How can something which does not exist (and may never exist) be the cause of a presently occurrent event? When we are dealing with conscious processes, we may evade this issue by regarding the *desire* for the accomplishment of the end as the causal factor which initiates the action. But, as we have seen, purposive behavior occurs even in the absence of conscious desires. The anticipatory pleasure of eating cheese may prompt me to go to the refrigerator in search of cheese, even if there happens to be no cheese there. But human beings tend to behave purposefully, for example, in acquiring an education, even where there is no clearly specifiable goal to be accomplished. Avoiding the language of consciousness and attributing some other kind of attractiveness to the goal will not do either, for, as we have indicated, behavior may be recognized as purposive even where the goal is not achieved or where the aimed-for goal does not exist.

The philosopher is concerned with the logic of teleological explanations. He would like to know whether all instances of goal-directed or purposive behavior can be accounted for according to the same model, and whether this model, in turn, is explicable in ordinary physical causal terms. He is not primarily interested in the kinds of things that exhibit teleological behavior. However, if it were the case that only living organisms do exhibit teleological behavior and that such behavior could not be explicated in terms of the ordinary physical laws which explicate action and reaction, then this would be evidence in favor of vitalism. Even if purposiveness were merely a characteristic life feature, typically found in living organisms, but also to be found in nonliving objects under special circumstances, as, for example, when they are the product of human design, then we might regard vitalism as somehow substantiated.

Increasingly, however, the development of inanimate objects with sophisticated ability to modify goal-directed behavior has led to a decline in the apparent necessity of maintaining vitalism. The situation is very much like that which prevails with respect to the origin of life. It cannot be demonstrated that life did arise in one or another specific fashion, but if a plausible case can be made for a naturalistic method by which it might have arisen, then there is no reason to assume anything more mystical. The situation with respect to teleological explanation is parallel. If apparently purposive behavior *can* be exhaustively accounted for in terms of an ordinary causal model, then there is no need to introduce an explanatory model which singles out such behavior as unique. And then there is no need to attribute a unique mode of existence to those objects which exhibit that type of behavior. But no conclusive proof can be given that living things lack uniqueness.

It was reasoning of this type which led some of the early systems theorists, such as Ludwig von Bertalanffy, to believe that they could by-pass the traditional mechanism-vitalism controversy altogether. Their aim was to introduce a new science of matter which would analyze it in terms of systems and of systems-of-systems of varying degrees of complexity. Living organisms would not require a novel approach, which treated them as entities uniquely differentiated from ordinary material objects because of their purposive behavior. Such behavior could be accounted for within the context of familiar causal principles. Organisms could be viewed as one form of the systemic organization of matter, distinguishable from other systems by their particular mode of organization, which is considerably more complex than other modes, but not fundamentally discontinuous with them.

This view permits an analysis of a *dead* organism, as distinct from a *living* organism without confusing it with an inanimate object, for the term death now refers to a systemic dysfunction rather than to a state of being. A dead organism is a disrupted system, analogous to a broken machine. Just as a machine, whose functioning depends upon its being "wound up," may be "run down," so a material system which, in its operative condition, is known as living, is, when inoperative beyond the point of repair, known as dead. This is to say that death does not represent a specific mode of malfunctioning, or a fixed degree of disruption, but rather that it is an expression used to indicate that the system is more than trivially damaged, and that there is no expectation of its rehabilitation. This interpretation of death is entirely consistent with the situation noted above—that the demarcation between life and death is an arbitrary one which shifts with our technical capacity for rehabilitation. It is also consistent with the claim that death is not localizable as the cessation of operation of any particular part or set of parts of the organism, but refers to the dysfunction of the whole system while some of the subsystems may remain functionally intact. Thus, to refer to something as dead is to indicate what kind of system it is and, further, to say that it is inoperative. The language that we use to identify systems and their properties is conventional. A system is a system by virtue of our designating it as such, not as a result of some inalienable natural law. We may view some systems, such as that of kinship relations as more "natural" than others (*e.g.,* crime syndicates). But this is only to say that some systems are based upon relationships which would exist apart from the system, while others postulate relations which exist only within that system and as related by it. Organisms would produce offspring regardless of our identification of families, but an international conspiracy exists only within the context of what society chooses to identify as national and as conspiratorial.

Analogously, we may think of death as referring to a systemic relation—specifically as denoting a profound breakdown within a systemic complex. Indeed the statement *"X* is dead" conveys the information that a system of a particular kind—one that was properly designated as living—is now dysfunctional.

While the term dead reflects dysfunction of a system which, if functional, would be properly qualified as living, the terms animate (or living) and inanimate refer rather to the type of system in question. Only an animate system has death as a characteristic mode of dysfunction. The terms dead and inanimate then cannot have identical meanings, but they have a clear logical connection. A system may be animate or inanimate, and only if it is animate, might it be dead.

The relation between the animate and inanimate, and the continuity between them may be raised in another context. Our discussion of biopoesis suggested that historically the living did arise out of nonliving matter. Furthermore, we know that living organisms do not contain any substances which are not also to be found in inanimate nature, and that if the process of spontaneous generation does not occur continuously, it is only because of external impediments which destroy the combinations that are formed. Recent successes in synthesizing DNA, and even viruses from artificially synthesized chains of DNA, give additional support to the thesis of the continuity of the inanimate and the animate. Indeed, we have seen how difficult it is to designate a fixed point which divides the living from the nonliving. The unit of life itself is a matter of dispute.

Before the discovery of the viruses, there appeared to be a break, at least in terms of size and structure, between the living and the nonliving. But the investigations of Wendell Stanley and others during the 1930s revealed that some viruses were smaller than known protein molecules and, in fact, were nucleoproteins— molecules consisting of protein and nucleic acid. They were crystallizable like inanimate protein molecules with which they shared the usual properties, and yet they possessed the capacity to mutate and to reproduce themselves or rather to divert the machinery of a host organism to the production of more viruses. Such characteristics were previously associated exclusively with living organisms. Viruses are now known to exist, filling the entire size gap between the well-defined animate and the inanimate. They are commonly regarded as filling a transitional phase between the living and the nonliving, but there is considerable debate as to whether they represent a survival of a primitive stage of life of an earlier evolutionary epoch or whether they are a degenerate form of a once more-sophisticated organism. One of the difficulties in reaching a conclusion on this matter relates to the fact that viruses now are dependent upon the existence of higher forms of life. While they have been artificially synthesized, they normally reproduce by entering, or rather by injecting the nucleic-acid portion of themselves into a host organism, and subverting the genetic mechanisms of the host so that it ceases to produce more of its own offspring, and instead produces more of the virus. This dependency suggests that the viruses could not have existed in their present form prior to the existence of other organisms more highly developed than themselves, but they might have undergone mutations which destroyed an earlier self-sufficiency.

Whichever the answer, the study of viruses has led to increasing certitude that the fundamental features of life are to be found in the nucleic acids, DNA and RNA, which are responsible for their own replication and for bringing about the replication and variation of other organisms. And once again, as the structure of the nucleic acids and the mechanisms of their behavior become increasingly understood and synthesizable, the gap between the liv-

ing and the nonliving narrows. Increasingly, this gap manifests itself as a function of human ignorance rather than as a breach
within the uniformity of nature.

We are often tempted to confuse the gaps in our understanding
with discontinuities in the order of nature. This tendency reflects a
fundamental philosophical problem concerning our ability to know
what is, and to transcend the limits of our own knowing apparatus. This problem is too broad in scope to be considered with any
hope of completeness here, and too profoundly important to be
neglected altogether. Consequently, it will be only briefly exposed
as a problem, with no serious attempt to elaborate a solution and
only some suggestions as to its relevance to the issue which is the
center of our attention.

Aristotle believed that by observation and study men could discover the "natural joints" of the universe. He believed that things
were essentially as common sense represents them, falling into
clearly definable classes of individuals, whose essences are fixed
once and for all and are intelligible to the rational mind. He conceived of the mind as a responsive instrument which registers the
nature of things upon itself and thereby knows them. The inventive capacity of the mind is, according to this view, minimal, and
the order of nature is wholly independent of it. It would have been
laughable to suggest that order is a creation of the human mind,
and in a later age, this suggestion would have been not merely
subject to ridicule, but to prosecution for blasphemy, if not for
heresy. Until quite recently there was no disputing the objectivity
of the world and the fundamental intelligibility of its nature. This
is the view of the nature of things and of our knowledge of them
which prevailed in the Western world for many centuries.

Subsequently, however, philosophical doubts began to be raised
about both the fixity and the objectivity of the orderly nature of
the universe. Philosophers became aware that more than beauty
may be in the eye of the beholder, and that beholders tend to disagree not only among themselves, but even with their own prior
judgments made in different contexts. They began to question the

passive role of the mind in the acquisition of knowledge, and they began to doubt the absoluteness of even those truths whose certainty had been represented as beyond rational doubt. Eventually, they probed the criteria by means of which we certify a proposition as true, and then they investigated the standards by which we justify those criteria. The outcome of all this was an increasing scepticism about the possibility of knowing anything at all and, specifically, of making true judgments about the nature of the universe. Total scepticism is rationally and practically impossible, for to claim that we cannot know is, implicitly, to affirm knowledge that we cannot know, or at least that we have a reliable way of repudiating knowledge, and this is self-contradictory. Furthermore, it is evident that we can and do make good judgments as well as bad ones, and that we can (retrospectively) and do differentiate between them. Consequently, we can substantiate, by providing reasons, our preference for understanding the universe according to one set of ordering principles as opposed to another. Although we are not in a position to claim that this or that set of principles truly describes nature, we may say that this or that set of principles is better than another competing set for the following reasons. It may, however, turn out that the same reasons do not always apply, and that an order which seems appropriate in one context is not suitable to another. It may be that a set of principles which appeared to account for the phenomena of experience fails to accord with an additional set of principles, or turns out to be less acceptable than another set of principles for entirely different reasons.

This is more or less the issue at stake with respect to teleological explanation. No one disputes that certain kinds of behavior can be described as goal directed, and that objects exhibiting such behavior can be classified as distinct from all other objects, requiring special modes of explanation. It may even be that there are advantages to be gained from making such a distinction. The question is, however, whether this is necessarily and under all circumstances the best way to understand those objects. While it is far from clear what constitutes the "best" way of understanding, and we cannot elaborate that point without going far beyond our initial intentions, we can, I think, agree that differences exist between explanatory models, and that some are more fruitful than others. There is no doubt that certain so-called "biological concepts" such as "gene" and "embryo" are useful in the formulation of biological explanations. The fact that genes and embryos can be literally isolated and broken down into their chemical constituents does not alter the fact that the unit concept, gene or embryo, is a valuable one for pursuing biological investigations. The point to be stressed here is that the judgment of what is fruitful and what is best is as much dependent upon ourselves as it is upon the nature of things apart from ourselves. Our selection of objects and of things and events to be explained is a reflection of our interests, rather than a wholly objective response to a nature which is independent of ourselves and which would be as it is regardless of our attention.

Returning then to the issue of natural discontinuities, as for example between the living and the nonliving, we can see that the crucial question is not whether or not there are such breaks. Whether there are or not, we cannot meaningfully establish, for any answer will be posed in terms of our own patterns of thought, the contexts of meaning we impose, and the decisions we make about what is the "best" way to view the world. The philosophical task of paramount importance is not the discovery of "what there is." Philosophy is not a form of superscience, nor is it in competition with the natural sciences. The philosopher must devote himself to the elaboration of those patterns of thought and contexts of meaning in terms of which we do structure the world. He must subject these patterns to constant scrutiny lest they become unconscious fixations which stultify our perception of the universe, rather than facilitate it. The carpenter who does not care for his tools soon begins to blame the wood for its refractoriness. Similarly, the philosopher who disregards the conceptual frameworks which are the tools of his trade eventually finds the objects of his inquiry obtuse and unresponsive. The issue of natural discontinuities, then, is not to be resolved in the laboratory by sheer observation.

Conceptual breaks are not like lesions or disconnected wires; they are matters for philosophical decision. Whether one is a mechanist or a vitalist is also a matter of philosophical choice. The dispute between mechanism and vitalism cannot be settled by empirical inquiry, for it rests upon a philosophical disagreement over the preferability of two incompatible modes of understanding the universe. The difficulty is compounded by the fact that disagreement revolves not only about the proper mode of understanding, but also about methods of justifying those modes of understanding; so that argument, even at this second level, is problematic. What is acceptable to one group as a reason for a conclusion is not acceptable, or may even be regarded as totally irrelevant or as sheer nonsense, by the other. The argument, insofar as argument is possible, runs at cross purposes. If arguments of this nature are allowed to flourish for long, eventually the antagonists become hardened in their positions, and begin discoursing only to their own true believers, and gradually develop separate languages which make further communication all but impossible.

It may well be that such an impasse has been reached with respect to an issue that lies at the heart of the controversy over natural discontinuities and thus at the root of the mechanism-vitalism dispute. This is the problem of reductionism. There is a tendency to think of reduction as a form of diminution. When you reduce something, it gets smaller; it loses impact. This is reflected in the expression that something is "nothing but a. . . . " The expression suggests that we have disposed of something which falsely appeared to be of greater significance than it turns out to be. Reducing something is like "putting it in its place," which often amounts to "putting down." While reduction is sometimes useful, especially where our objective is the achievement of clarity, the concept of reduction tends to retain associations of minimizing or depreciation. These associations are exhibited in the disposition to resist philosophical reduction, or the reductive explanation of a scientific theory in terms of other, more primary scientific laws and principles. But explanatory reduction is essentially the illumination of one set of principles by making them comprehensible in terms of another set of principles which are already understood. Nothing happens to the things or events to which those principles apply. They are surely not diminished.

There is an ideal of science, perhaps an ideal of all rationality, that all the variety of phenomena in the universe should be understandable in terms of one or a few fundamental principles. The endeavor to find such universal principles has preoccupied intelligent men throughout the history of Western civilization. This is the aim which motivates those who seek to reduce one science or set of scientific principles to another. Their objective is not to reduce or diminish the objects in the world, but only the explanations of them. Heat, for example, is explained by the kinetic theory of gases in terms of the mean kinetic energy of the gas molecules, and ultimately of the motion of particles. This does not constitute a denial of the existence of heat; it does not entail the illusoriness of the felt phenomenon of heat. It merely affirms that this phenomenon can be accounted for in terms of the same laws and principles which are applicable to motion of particles. Most so-called chemical qualities, qualitative descriptions of solids, liquids, and gases, are explicable in terms of accounts of sub-atomic structures, and it is generally conceded that the science of chemistry is reducible (though the reduction has not been completed) to the science of physics. In a complete reduction, the laws of the reduced science may, with the help of certain "rules of transformation" be deduced from the laws and principles of the reducing science.

Biological reduction would explicate those features which are characteristic of living things in terms of the laws and principles of physics and chemistry. It is, of course, understood that biological entities are also physical entities, and that consequently the laws of physics apply to them as such. A man falling out of a window is subject to gravitational attraction just as if he were a rock, and many physiological features of organisms can be simply accounted for in mechanical terms. The dispute turns upon certain characteristics and patterns of behavior which are not found in nonliving things, but which, according to reductionists, can nevertheless be explained in terms of the same physical principles which are employed to account for the phenomena of the nonliving world. Nonreductionists deny this explicability and declare that there are unique biological features which require the introduction of unique biological laws. They, in other words, do not deny that organisms are physical objects to which physical laws apply, but they believe that physical explanations arc insufficient to account for all the features of organisms. Biological objects, they say, are more complex than purely physical objects; and, while this does not necessarily imply that they violate the laws of physics—although some vitalists will maintain even that—it does require the extension of physical law. Thus, nonreductionists are constantly looking for new laws which apply exclusively to biological phenomena, or to some feature of biological phenomena.

It should be noted that even reductionists may grant the desirability of formulating biological laws. It may be practically unfeasible to express such things as genetic and developmental behavior in terms of physical laws, but that does not mean that it is logically impossible. In fact, as we have noted, it is evident that great strides have been made in biology through working with purely biological concepts. Such terms as "gene," "embryo," or even "organism" have no counterpart in a purely physical language, and it would surely be unreasonable for an ecologist or a marine biologist to devote himself to minute atomic analyses of his subject matter. It is sometimes argued that biological explanations, just as teleological description, are merely approaches to the same body of data as dealt with by physical theory, but from a different organizational and functional point of view. A body may be regarded "as organism" or "as a spatiotemporal entity." Reductionists are not advocating the second, nor are they trying to make physicists of all their fellow-scientists. They are not saying that rabbits and trees and ecological communities are "nothing but" a bunch of atoms. They do maintain that ultimately and ideally all processes, biological as well as physical, are expressible in terms of a single set of principles. The endeavor of molecular biologists to explain heredity and differentiation as an instance of the storage and transfer of information is a case in point. Their aim is not to affirm or deny the reality of the existence of anything, but to defend the sufficiency and universal applicability of a set of principles.

It is important to bear in mind in arguing this position that the sciences of physics and chemistry are not yet complete. Most reductionists do not mean to say that biological laws are reducible to today's physics, but that a complete physics will be sufficient in scope to include the principles of biology as well as those of non-living phenomena. But this is a claim that is open to a wide variety of interpretations, for what is "a complete physics"? Once again we meet the problem of the classification of the sciences. If we prejudge the issue by declaring physics to be the study of the inanimate, while biology is the study of the animate, then it follows of necessity, but also trivially, that no physics will be adequate to encompass biology. There is no fixed definition of the science of physics. If we declare it broad enough, it will include biology of necessity. Conceivably, the "ultimate" science from which all the laws of all the phenomena in the universe will be derivable will not be properly designated as physics or as any science with a label now familiar to us. Possibly reductionists are looking toward the wrong science.

Some philosophers and scientists have suggested that it is an error to seek to explain the complex in terms of the simple. Most of us have been indoctrinated with the geometric principle that complicated things are built out of simple ones, and that fruitful analysis is the result of careful separation of the simple elements which together produce a complex synthesis. We have adopted this principle as a pedagogic rule, and it has colored our view of the universe. But there are other modes of understanding. Sometimes we are able to understand a simple notion, not as a condition of, but as a consequence of our understanding of something more complex. That is, once we have a comprehension of a complex situation, individual features of it seem to fall into place with respect to it, but we could not have understood them apart from or prior to placing them within that total context. Perhaps the universe does not consist of simple elements combined into complex patterns, and perhaps, as a number of philosophers have suggested on the basis of a variety of philosophical grounds, we cannot hope to understand the part (the "simple") while lacking an understanding of the whole (the "complex"). On this alternative analysis of thought, it would follow that among the familiar sciences it is not biology which is to be reduced to physics, but the reverse; furthermore, biology would presumably be reducible to a science such as psychology, which would be reducible to one of the social sciences. I do not mean to advocate this reverse reductionism, but only to indicate that there is nothing sacrosanct about the pattern of reductionism which is most ardently endorsed by contemporary reductionists. It may well be that one day we will scrap the entire classification system of the sciences, which is our heritage, and begin anew with a reexamination of phenomena according to a fresh set of categories.

The suggestion that the relations among the sciences and between the sciences and the world need not be inflexibly cast raises another philosophical issue which has relevance to biological reduction. That is the knotty problem of determinism. One of the primary reasons reductionism produces anxiety in its opponents is because they hold that, if biology is not independent of physics, then it is reduced to a deterministic science.

The claim that biology is an autonomous science or independent of other sciences is intended, in part, to deny that biological events are under the constraint of physical law. But let us consider what is meant by this denial. It is clear, as we have seen, that a man falling out of the window is subject to the principle of gravitation. He is a physical object, and, as such, he is not significantly different from a stone or flowerpot whose trajectory is causally determined by identical factors. But a distinction must be made between the independence of an event as free of causal determination and the independence of a scientific law or principle as nonderivable from other laws and theories of science. The defense of biological autonomy on the theoretical level is meant to reject the total subsumption of the laws of biology under the laws of physics. It is often mistaken to be a denial that biological events (whatever these might be) are physically determined.

To claim that one science is autonomous of another is to say that they are logically independent. The laws of one may not be deduced from those of the other. This is not to say that phenomena which are explainable in terms of one set of laws might not also be explainable in terms of the other. But the two explanations would be complementary, focusing upon two different aspects of the event in question, neither explanation being sufficient to accommodate all features of the situation at once. The independent sets of laws are just that—independent. They neither entail one another, nor do they preclude one another. The autonomy of the laws is with respect to one another, not to events, for a given event may be determined with respect to other events, or to laws or to both events and laws.

The notion that an *event* is autonomous is almost rationally incomprehensible, for what it implies is that the event is *uncaused,* that it is causally unrelated to any prior or simultaneous set of events (not theories or laws). Extreme indeterminists—and they are usually irrationalists—maintain that there are such events, *gratuitous acts,* which cannot be causally explained at all. Such acts do figure in works of existentialist literature. The unmotivated, unpremeditated murder is a favorite topic (*e.g.,* of Dostoevski in *Crime and Punishment* and of Gide in *Lafcadio's Adventures*), but it is highly problematic and is invariably accounted for as caused after all in terms of some psychological or psychosocial explanatory framework. It actually is extremely difficult to conceive of, much less talk about, an act as wholly fortuitous. But that, strictly speaking, is what is meant by declaring an event autonomous.

More commonly, when an author refers to an event as autonomous, what he has in mind is its inexplicability in terms of a *particular* causal theory. But in this sense all events are autonomous with respect to some theories. The falling of a star cannot be explained on psychological grounds, just as the fluctuation of fashions in women's hemlines cannot be explained in terms of the principles of physics or of microbiology. But this does not mean that astronomical events or the current craze for miniskirts are uncaused and cannot be explained at all. There is no single theory that will account for every event in the universe. The suggestion that "not a sparrow falls without God's knowing it" is not a counterargument, for it offers no explanation of such events.

All events, however, at least those that can be identified and meaningfully discussed, can be explicated in terms of some theory or causal account. The belief that every event has a cause is one of our most fundamental convictions, and we do not give it up, even when we are ignorant of the specific cause of a specific event. We simply look harder and more imaginatively.

Those who affirm that biology is autonomous, then, certainly do not mean to say that biological events are uncaused. They mean that the laws and principles of biology cannot be "reduced to" or derived from the laws and principles of another science, such as physics. Furthermore, they do not mean that biological events are not governed by necessity. The issue between autonomists and nonautonomists is not concerned with constraint. Nor does determinism deny that things could be other than they are. It is sometimes believed that the denial of determinism is intended to uphold the conviction that there are some events, notably human volition or acts of will, which are not subject to necessity but are free. By contrast, determinists are represented as maintaining that all actions, human volitions included, are constrained by necessary laws. But this is to confuse determinism with fatalism. The fatalist contends that things could not be other than they are, that all events follow an invariable and irrevocable law. The determinist says that all events are caused; they could be otherwise, but this would require manipulation of their causal antecedents. In other words, they are not *necessarily* as they are, but they are *predictably* so. The possibility of prediction does not impose any constraint upon events; but is, if anything, itself constrained by the causal order among events. We can make predictions because there is order. We do not create order by making predictions.

The order that exists, however, is not a necessary one. It could be other than it is; our description of it and predictions about it might be mistaken. The possibility that prediction might be in error or even impossible has spurred the indeterminists into making new claims. Traditionally it was believed that only events in the inanimate world were predictable, since they were governed by inexorable law. But human actions, or at least some of them were hailed as nondeterministic. One of the characteristic features of life, it was said, is this very unpredictability. The sheer improbability of the occurrence of an event as complex as life seemed to reinforce the claims of the indeterminists.

But the physics of the twentieth century introduces indeterminacy at the ultimate level of subatomic particles. No longer is it believed that the realm of inanimate objects is subject to rigid, deterministic law, while freedom prevails within the domain of life. On the contrary, the principle of uncertainty refers to the impossibility of simultaneous description of the velocity and position of a single electron relative to its nucleus. At the biological level, which is macromolecular, the number of subatomic interreactions which occurs is so great that they cannot be individually considered. And, hence, calculating statistically, the indeterminacy at the individual level is canceled out. At the macromolecular level, causal laws can be formulated. This would imply that the converse of the classical vitalistic principle is correct—that indeterminacy prevails at the micro level of inanimate phenomena, but that at the animate level large-scale events, which are governed by statistical laws, are determinate.

But it would be an error to claim that our new ideas about determinism mark a triumph for mechanism. The fact is that the issue between determinism and indeterminism is no more central to the mechanism-vitalism dispute than is the issue between materialism and its denial. It is true that if it were demonstrable that *either* inanimate *or* animate phenomena *exclusively* were indeterminate, then a ground for radical discrimnation would exist, and there would be some justification for declaring the truth of vitalism. Note that it would be immaterial whether indeterminism prevailed in the animate or the inanimate domain. Neither mechanism nor vitalism is committed to any partisan position on the determinism-indeterminism issue. There have always been vitalistic theories which are deterministic. The most extreme form of vitalism, based upon divine decree, is a case in point. But any vitalistic theory which views life as subject to its own laws, absolute, but distinct from the laws which govern the physical world, is bound to be deterministic. Dualism does not preclude determinism; it only affirms two sets of determining laws, working in tandem. Mechanism, on the other hand, need not be and at present is not deterministic. The mechanist is limited only to maintaining that whatever principles apply to the universe as a whole, be they deterministic or indeterministic, or possibly developmental, they are sufficient to explicate the animate as well as the inanimate realm.

In order to deal with the mechanism-vitalism controversy, then, we need not resolve the issue between determinism and indeterminism. While the problems are not unrelated, the solution of one is not dependent upon the solution of the other. It may be remarked in passing that belief in determinism has had considerable rewards in the promotion of scientific research. While no one has or is ever likely to demonstrate the truth of determinism, it is intellectually productive to assume that nature is orderly, that events may be related to one another in terms of causal laws, and that the laws which describe patterns of events can, in turn, be incorporated in explanatory theories. Were we to assume that some events have no causes or are inexplicable, we would not know when to stop looking, and we would be inclined to give up in the face of the mildest frustration. Assuming that an explanation is to be found, we are motivated to search for it. Thus, while the assumption may be, at this point, immodest, there is ample practical justification, if no other, for making it.

The resistance against determinism in human affairs and in biological matters, just as the inclination to reject mechanism, appear to be residual features of a basically antirationalistic philosophical position which regards knowledge as a sometimes-objectionable thing. Perhaps most of us encounter occasions when we would prefer not to know something. The person with an incurable disease may be happier with an uncertain illusion than with the certainty of his fate. Some scientists have explicitly expressed their unwillingness to press their research beyond certain questions, simply because they admit possibilities which they prefer not to understand. All of us have had the experience of "knowing too much." Recently, scientists have expressed concern about the possible political and moral implications of the isolation of the gene. However satisfying such knowledge may be on purely intellectual grounds, the possible uses to which it might be put eugenically are terrifying. For many people, the very mystery of the unknown is an attraction, and that which is understood is devalued in proportion to the decline of mystery. That which is understood no longer holds any fascination and may even be dismissed as lacking in interest. Conversely, where we are conditioned to regard something with reverence and to attribute a high value to it, there we may also encounter reluctance to understand it too well. However one might deplore such resistance as intellectual cowardice or even dishonesty, it is a familiar occurrence, and it is not surprising that life itself should be one of those taboo areas which we are warned against probing too deeply.

But there is something of word magic in this, for to call a thing by name is not to deprive it of its powers. Supposing that we were to convince ourselves that life was a natural phase of the development of matter; that given certain antecedent conditions, it was indeed inevitable that life, in more or less the form that we know it, should occur. Would this make the phenomenon of life any the less remarkable or worthy of our admiration? I fail to see any reason why that which is understood should thereby stand diminished in the eyes of its beholder. On the contrary, we frequently have all the greater respect for something whose complexity we are capable of recognizing.

Among the most pronounced instances where appreciation is enriched by understanding are aesthetic responsiveness and interpersonal relations. While it is possible to respond to a work of art or a beautiful object strictly in terms of one's visceral pleasure ("I know what I like, but I don't know why"), the gratification is surely more sustained and possibly more intense if one does understand what it is that one is contemplating. The connoisseur of baseball, or of bullfighting, or of any game, sees the plays that other people miss, and his enjoyment is of greater depth. Similarly, in personal relationships, our attachment to people grows as we come to know them better, and those whom we love and hold in friendship are those whose characters seem to us most worth the effort of understanding. It is simply not true that love must be blind and that we can only revere that of which we are ignorant, though this is a popular myth.

All this is not necessarily to say that mechanism is the way of rational enlightenment, while vitalism is the last stronghold of obscurantist mystery mongers. To a large extent, it is the very challenge of vitalistic questions and counter solutions which keep mechanists in pursuit of answers. And just as constructive criticism is necessary from within the framework of science itself, so the whole enterprise must constantly be exposed to external review and to the frequently hostile criticism of individuals with other values and other aims.

The more we come to understand about the nature and origin of life, the more we appreciate the immensity of our ignorance about it. There is increasing evidence favoring a spontaneous and natural origin of life and development of living things in a manner consistent with our knowledge of the physical world. But while our conjectures can be made consistent with the body of knowledge already within our possession, there is no way of knowing what we do not and cannot know. Ultimately, what we believe is a matter of personal commitment, but as intelligent human beings, we have an obligation to investigate even those commitments.

About the Author

DR. HILDE HEIN *received her doctorate in philosophy at the University of Michigan, where she subsequently taught. She has since taught at Los Angeles State College and Tufts University, and is now Associate Professor of Biology at College of the Holy Cross, Worcester, Massachusetts. She is the author of numerous articles and is now writing a major work on vitalism and mechanism.*